User-Centric and Information-Centric Networking and Services

User-Centric and Information-Centric Networking and Services

Access Networks, Storage and Cloud Perspective

Edited by

M. Bala Krishna

CRC Press
Taylor & Francis Group
Boca Raton London New York

CRC Press is an imprint of the
Taylor & Francis Group, an **informa** business

CRC Press
Taylor & Francis Group
6000 Broken Sound Parkway NW, Suite 300
Boca Raton, FL 33487-2742

Library of Congress Cataloging-in-Publication Data

Names: Krishna, M. Bala, author.
Title: User-centric and information-centric networking and services : access networks, storage and cloud perspective / M. Bala Krishna.
Description: Boca Raton : CRC Press, Taylor & Francis Group, 2019. | Includes bibliographical references.
Identifiers: LCCN 2018046536 | ISBN 9781138633322 (hb ; alk. paper) | ISBN 9781315207650 (ebook)
Subjects: LCSH: Computer networks. | Information networks.
Classification: LCC TK5105.52 .K75 2019 | DDC 004.6--dc23
LC record available at https://lccn.loc.gov/2018046536

Visit the Taylor & Francis Web site at
http://www.taylorandfrancis.com

and the CRC Press Web site at
http://www.crcpress.com

Contents

List of Figures

List of Tables

Preface

User-centric networking (UCN) extensively uses centralized and distributed systems. With extensive research and advancements in information-centric networking, various user-centric systems are being evolved and functionally adapted in homogeneous and heterogeneous systems. UCN is established and controlled by a single user or a set of users in the network. User-centric nodes are selected based on attributes such as node profile, location, connectivity factor, buffer size, network processing speed, and traffic conditions. The connection models such as fixed user-centric model and location-based user-centric model enhance the features of centric-based service applications. UCN aware push service uses the content delivery architecture and pre-cached features for multimedia services in mobile devices and efficiently uses the channel bandwidth.

Information-centric networking (ICN) is modeled using information, priority, and application service requirements. ICN applies the data aggregation and data fusion methods to collect the data and forwards it to the data storage node or service node. Security layer encrypts the private and confidential node information in the network. ICN is classified as follows:

- Content-centric networking (CCN): Network established based on the type of information and priority generated by the user nodes
- Service-centric networking (SCN): Network established based on the set of services that include the information, query, services, device type, and so on

This book provides state of the art and novel approaches for user-centric and information-centric networking and services, covering current and emerging topics, so that it can be an excellent reference book for developers, researchers, academicians, and graduate students. This book comprises contributions from prominent researchers and academicians working in the area of user-centric networks, information-centric networks, storage-centric cloud, and access networks. This book aims to enlighten readers in the emerging areas of user-centric and information-centric networking.

Chapter 1 explains the features of centric-based networking systems (CNS) that operate in centralized and distributed modes, enhance the content delivery rate, and reduce the operational and economic cost of the system. This chapter classifies the CNS based on parameters such as user preference, information type, content type (centralized or distributed), address range, network type, network modes (device-centric or address-centric), data factor (aggregation or dissemination), and service types.

Chapter 2 explains the naming and addressing schemes used in ICN. Naming approaches used in routing, security, and privacy are explained in this chapter. Various naming schemes such as flat naming, hierarchical naming, and hybrid naming based on primary and secondary parameters are explained in this chapter. Naming and addressing schemes in the mobile scenario are explained in this chapter.

Chapter 3 explains the information-centric exchange mechanisms for IoT interoperable deployment. Named publish subscribe networking with named data and protocol, ICN exchange mechanisms based on architecture, address translation, signaling adaptation, and content adaptation are described in this chapter. IoT interoperable deployment, scalability, and management are explained in this chapter.

Chapter 4 explains the information-centric networking for future internet video delivery. Internet video streaming requirements based on demand, live, and dynamic adaptive streaming and evolution of video transmission techniques are highlighted in this chapter. IoT video transport, SDN for information-centric video distribution, policies, and regulations that affect ICN are explained in this chapter.

Chapter 5 explains the classification of storage-centric networking. Attributes, models, and hierarchy of storage systems are explained in this chapter. Storage-centric systems based on thermal power, differentiated services, load management, and compressive sensing are described in this chapter. Storage virtualization, security, and privacy for SDN and wireless networks are explained in this chapter.

Chapter 6 explains the cloud-centric networking using virtualization and resource sharing. Multicloud using optimal network function is highlighted in this chapter. Virtualization operations and enhancements for supported services are explained in this chapter. Resource sharing based on invitation, shared aspects, and centralized and distributed approaches are explained in this chapter. Security challenges based on SDN and NFV are elucidated in this chapter.

I take this opportunity to express my sincere gratitude to all authors, co-authors, and the publishing team. Firstly, I am deeply indebted to the active authors and researchers, who shared the vision of this book "User-Centric and Information-Centric Networking and Services: Access Networks, Storage and Cloud Perspective" and contributed to high-quality chapters. I sincerely thank authors for their tireless effort in bringing this book to a noble and presentable form. Secondly, I thank various technical societies for their kind support. Thirdly, I thank the CRC Press editorial team, especially Richard O'Hanley, Stephanie Place, Rachael Panthier, and Kiruthiga Sowndar for their patience, guidance, support, and encouragement during the initial, progressive, and final phase of this book. Finally, I thank my family members for their warm support and encouragement.

M. Bala Krishna
University School of Information and Communication Technology
(USICT)
Guru Gobind Singh Indraprastha University
New Delhi, India

Editor

M. Bala Krishna (Bala Krishna Maddali) earned his Bachelor of Engineering (BE) degree in Computer Engineering from Delhi Institute of Technology (presently Netaji Subhas University of Technology, and formerly Netaji Subhas Institute of Technology), University of Delhi, Delhi, India, and Master of Technology (MTech) degree in Information Technology from University School of Information Technology (presently University School of Information and Communication Technology), Guru Gobind Singh Indraprastha University, Delhi, India. He had earned Doctor of Philosophy (PhD) degree in Computer Engineering from JMI Central University, New Delhi, India. He had earlier worked as senior research associate and project associate in Indian Institute Technology, Delhi, India, in the areas of digital systems and embedded systems. He had worked as faculty member and had handled projects related to networking and communication. He is presently working as Assistant Professor in University School of Information and Communication Technology, Guru Gobind Singh Indraprastha University, New Delhi, India. His areas of interest include computer networks, wireless networks, advances in mobile computing and communications, cognitive radio and software-defined networks, security in communication networks, and the Internet of Things. He has publications in international journals, conferences, and book chapters. He has contributed as the corresponding co-editor for an international book titled as *Advances in Mobile Computing and Communications: Perspectives and Emerging Trends in 5G Networks*. He is an IEEE senior member and has served as the Workshop Co-Chair and Tutorial Chair/Co-Chair for IEEE conferences. He has organized the IEEE and ACM workshops. He is a Technical Program Committee (TPC) member for IEEE conferences.

Contributors

M. Bala Krishna
University School of Information
    and Communication
    Technology
Guru Gobind Singh
    Indraprastha University
New Delhi, India

Daniel Corujo
Department of Electronics,
    Telecommunications and
    Information Technology
University of Aveiro
Aveiro, Portugal

Antonio Skarmeta
Department of Information and
    Communications Engineering
Computer Science Faculty
University of Murcia
Murcia, Spain

Carlos Guimarães
Department of Electronics,
    Telecommunications and
    Information Technology
University of Aveiro
Aveiro, Portugal

Diogo Gomes
Instituto de Telecomunicações
Universidade de Aveiro
Aveiro, Portugal

G. Akhil
Cisco Systems Inc.
Bangalore, India

Helder Moreira
Instituto de Telecomunicações
Universidade de Aveiro
Aveiro, Portugal

João Paulo Barraca
Instituto de Telecomunicações
Universidade de Aveiro
Aveiro, Portugal

José Quevedo
Departamento de Eletrónica,
   Telecomunicações e
   Informática
Universidade de Aveiro
Aveiro, Portugal

Jordi Ortiz
University of Murcia
Murcia, Spain

Pedro Martinez-Julia
National Institute of Information
   and Communications
   Technology (NICT)
Tokyo, Japan

Rui Ferreira
Department of Electronics,
   Telecommunications and
   Information Technology
University of Aveiro
Aveiro, Portugal

Rui L. Aguiar
Department of Electronics,
   Telecommunications
and Information Technology
University of Aveiro,
Aveiro, Portugal

Samar Shailendra
TCS Research & Innovation
Bangalore, India

Contributors Biography

M. Bala Krishna (Bala Krishna Maddali) earned his Bachelor of Engineering (BE) degree in Computer Engineering from Delhi Institute of Technology (presently Netaji Subhas University of Technology, and formerly Netaji Subhas Institute of Technology), University of Delhi, Delhi, India, and Master of Technology (MTech) degree in Information Technology from University School of Information Technology (presently University School of Information and Communication Technology), Guru Gobind Singh Indraprastha University, Delhi, India. He had earned Doctor of Philosophy (PhD) degree in Computer Engineering from JMI Central University, New Delhi, India. He had earlier worked as senior research associate and project associate in Indian Institute Technology, Delhi, India, in the areas of digital systems and embedded systems. He had worked as faculty member and had handled projects related to networking and communication. He is presently working as Assistant Professor in University School of Information and Communication Technology, Guru Gobind Singh Indraprastha University, New Delhi, India. His areas of interest include computer networks, wireless networks, advances in mobile computing and communications, cognitive radio and software-defined networks, security in communication networks, and the Internet of Things. He has publications in international journals, conferences, and book chapters. He has contributed as the corresponding co-editor for an international book titled as *Advances in Mobile Computing and Communications: Perspectives and Emerging Trends in 5G Networks*. He is an IEEE senior member and has served as the Workshop Co-Chair and Tutorial Chair/Co-Chair for IEEE conferences. He has organized the IEEE and ACM workshops. He is a Technical Program Committee (TPC) member for IEEE conferences.

Daniel Corujo is a doctoral researcher at the University of Aveiro, Portugal, where he received his PhD in 2013, and a group coordinator at the Telecommunications Institute, in Aveiro, where he is coordinating the telecommunications and networking group. He has fulfilled different research and coordination roles in several national academic and industrial projects, as well as international ones, under the scope of the EU FP7, such as DAIDALOS, OneLab2, 4WARD, MEDIEVAL and OFELIA. He has over 70 published conference and journal articles, and is a regular TPC member in several conferences and magazines. He has multiple contributions to different standardization organizations, such as the Institute of Electrical and Electronics Engineers (IEEE) and the Internet Engineering Task Force (IETF). He is author of the first IETF RFC on ICN (RFC 7476), and of two upcoming RFC's of the ICN Research Group. He was, until 2017, secretary of the IEEE ComSoc Portugal chapter. He lectures network and computer science classes and coordinates several MSc and PhD theses at the Universidade de Aveiro. He was also an adjunct lecturer at the Escola Superior de Tecnologia e Gestão de Águeda, in Águeda, Portugal.

Antonio Skarmeta received an MS degree in computer science from the University of Granada and a BS (with honors) and PhD in computer science from the University of Murcia, Spain. Since 2009, he has been full professor in the same department and university. Skarmeta has worked on different research projects in national and international networking, security and Internet of Things (IoT), participating in research projects like Euro6IX, ENABLE, DAIDALOS, SWIFT, SEMIRAMIS, SMARTIE, SOCIOTAL and IoT6. His main interests are the integration of security services, identity, IoT and Smart Cities. He has been head of the research group ANTS since its creation on 1995. He is also advisor to the vice-rector of research at the University of Murcia for International Projects and head of the International Research Project Office. Since 2014, he has been Spanish national representative for the MSCA within H2020. He has published over 200 international papers and is a member of several program committees. He has also participated in several standardization projects for the Internet Engineering Task Force (IETF), the International Organization for Standards (ISO) and the European Telecommunications Standards Institute (ETSI).

Carlos Guimarães received his computer and telematics engineering MSc in 2011 from the electronics, telecommunication and informatics department at the University of Aveiro, Portugal. He is currently pursuing his PhD degree in computer science of the Universities of Minho, Aveiro, and Porto (MAPi), in the area of future internet architectures. More specifically, he has been focusing on the study of interoperability mechanisms between future internet architectures (such as information-centric networks) and the current internet architecture, as a way to promote the initial roll-out of such architectures. Since 2010, he has also been a researcher and developer in the Telecommunications and Networks–Av group at Instituto de Telecomunicações–Pólo de Aveiro, Portugal (IT-Aveiro), working in the areas of software-defined networks, information-centric networks, mobility management and media-independent handover mechanisms. He participated in EU FP7 MEDIEVAL and EU FP7 OFELIA. He has also contributed to IT-Aveiro internal projects such as ODTONE and AMazING.

Diogo Gomes graduated with a degree in computers and telematics engineering from the University of Aveiro in 2003 with first class honors and received a PhD at the same university, focusing on resource optimization for broadcast networks in 2009. He's currently an auxiliary professor at the University of Aveiro. Over the last 10 years, has participated in several EU-funded projects such as IST-Mobydick, IST-Daidalos, IST-Akogrimo, IST-C-MOBILE, ICT-C-Cast, ICT-Onelab2 and ICT-Medieval, where besides conducting research on QoS, IP mobility, multicast/broadcast and service and application development, he has been deeply involved in the deployment of prototypes and demonstrations. His recent research interests are related to knowledge extraction and context storage in Internet of Things (IoT) scenarios using machine learning techniques and big data repositories.

G. Akhil is currently working as a network consultant at Cisco Systems, Inc. He has obtained his bachelor's and master's degrees from International Institute of Information Technology–Bangalore (IIIT-B). Akhil has worked with organizations such as Nokia and Cisco on 5G research. His current areas of research include ICN, 5G, IOT, SDN and segment routing.

Helder Moreira graduated with a degree in computers and telematics engineering from the electronics, telecommunications and informatics department at the University of Aveiro in 2016. His MSc dissertation was developed under the Smart Lighting project at Instituto de Telecomunicações, on sensor data integration and management of smart environments, under the supervision of Professor João Paulo Barraca and engineer Nuno Lourenço, from Think Control. After graduation, he worked at ThinkControl for several months, and later joined the University of Aveiro as a researcher, where he is currently participating in life-PAYT and smart green homes (SGH) projects. Additionally, he has also started doctoral studies, following his main interests in security, privacy and the Internet of Things (IoT).

João Paulo Barraca received a PhD degree in informatics engineering in 2012 from the University of Aveiro, where he developed work focused on the integration of social structures into self-management network functions. He is currently an invited assistant professor at the University of Aveiro, in areas related to programming, networking and security. Most of his research is conducted through Instituto de Telecomunicações (IT), with which he has been associated since 2003, and he acts as coordinator of the Aveiro Telecommunications and Networking Group (ATNoG). During this time at IT, he has published works in the areas of networking and software for computer systems. He has also acted as reviewer for several conference papers and journals, mostly international. He has participated extensively in both nationally and internationally funded projects, including FP7 IST Daidalos, IST WIP, IST S(o)OS, IST-Onelab, IST

Prose, CAPES DEVNF and FCT CRUISE. He is also a member of the telescope manager consortium for the Square Kilometre Array (SKA) telescope, overseeing tasks related to the cloud computation infrastructure (LINFRA) for the software management components.

José Quevedo graduated in 2009 from the Havana University of Technologies José Antonio Echeverría (CUJAE), Havana, Cuba, where he studied telecommunications and electronics engineering. Upon graduation, he was a junior professor at the CUJAE, where he also acted as information security officer of network infrastructure and worked. He is currently pursuing his PhD at the MAP-tele doctoral program in telecommunications at Universidade de Aveiro, Aveiro, Portugal, and also working as a researcher in the Telecommunications and Networks–Av group at Instituto de Telecomunicações–Pólo de Aveiro (IT-Aveiro). His current research focuses on information-centric networking (ICN) approaches for supporting Internet of Things (IoT) scenarios. His work has been disseminated in different publications tackling the challenges of exposing ICN to the complexity commonly associated with IoT scenarios, ranging from discovery to interoperability and mobility. He has also supported IT-Aveiro projects such as SeLF-ICN and AMAziNG.

Jordi Ortiz received a BS (2008) and an MSc (2009) in computer science from the University of Murcia, Spain. Since 2007, Ortiz has been a full-time researcher associated with international projects such as DAIDALOS, SCALNET, SMARTFIRE, OPENLAB, GEANT 3 & 4, STORK2 and ANASTACIA among others. His main fields of interest are identity federation, video streaming, Software Defined Network (SDN) networks and Internet of Things (IoT) networks. He has published international papers and serves as technical program committee member at some conferences and for journals.

Pedro Martinez-Julia received a BS in computer science from the Open University of Catalonia. As well, he received an MS in advanced information technology and telematics and a PhD in computer science from the University of Murcia, Spain. He is currently a full-time researcher with the National Institute of Information and Communications Technology (NICT), Tokyo. He has been involved in EU-funded research projects since 2009, leading several tasks and activities, and participating in the Internet Engineering Task Force (IETF) and the Internet Research Task Force (IRTF) for the standardization of new network technologies. He has published over 20 papers in peer-reviewed conferences and journals. His main expertise is in network architecture, control and management, with a particular interest in overlay networks and distributed systems and services. He is a member of the Association for Computing Machinery (ACM) and the Institute of Electrical and Electronics Engineers (IEEE)

Rui Ferreira received his computer and telematics engineering MSc in 2008 from the electronics, telecommunication and informatics department, University of Aveiro, Portugal. Since 2008, he has been a researcher in the Telecommunications and Networks group at Instituto de Telecomunicações, carrying out research work in the areas of network privacy, identity management, and security. As part of his work, he participated in EU-funded projects FP7 SWIFT and FP7 PROSE, as well as other internal research and development projects in associated areas. Currently he is pursuing a PhD in the topic of security and privacy in name resolution systems, studying the security and privacy implications of name assignment in computer networks at the various layers of the network stack and introducing techniques for enhancing name resolution systems.

Rui L. Aguiar received his degree in telecommunication engineering in 1990 and his PhD in electrical engineering in 2001 from the University of Aveiro. He is currently a full professor at the University of Aveiro, teaching in the networking area, and was previously an adjunct professor at the INI, Carnegie Mellon University. He was a visiting research scholar at Universidade Federal de Uberlândia, Brazil, for three years. He is coordinating research nation-wide at the Instituto de Telecomunicações, in the areas of networks and multimedia. He is leading the Technological Platform on Connected Communities, a regional cross-disciplinary, industry-oriented activity on smart environments. His current research interests are centered on the implementation of advanced wireless networks and systems, with special emphasis on 5G networks and the Future Internet. He has more than

450 published papers in those areas, including standardization contributions to Institute of Electrical and Electronics Engineers (IEEE) and Internet Engineering Task Force (IETF). He has served as technical and general chair of several conferences, from IEEE to Association of Computing Machinery (ACM) and International Federation for Information Processing (IFIP) and is regularly invited for keynotes on 5G and future internet networks. He sits on the TPC of all major IEEE ComSoc conferences. He participated extensively in national and international projects, of which the best example is his position as chief architect of the IST Daidalos project, and he has extensive participation in industry technology transfer actions. He is currently associated with the 5G Infrastructure Public Private Partnership (5G PPP) Infrastructure Association and is the current chair of the steering board of the Networld2020 European Technology Platform (ETP). He is senior member of IEEE, Portugal ComSoc Chapter Chair, and a member of ACM. He is associate editor of Wiley's Emerging Telecommunications Technologies (ETT) and Springers' Wireless Networks journals and has helped with the launch of Elsevier's ICT (Information & Communications Technology) Express. He is a chartered engineer and has acted as a consultant to several major operators, as a technology advisor to bootstrap several SMEs, and as an expert serving several public bodies, both on the societal and on the judiciary branches. He currently sits on the advisory board of several EU projects and research units.

Samar Shailendra is currently working as a scientist at TCS Research and Innovation. He is also a visiting faculty member at Indian Institutes of Information Technology (IIIT), Bangalore. He has more than 10 years of industry and academic experience. He has obtained his MTech and PhD from Indian Institutes of Technology (Delhi and Guwahati, respectively). Dr. Samar has worked with various research and software industries in the past such as Novell Software and Infosys Technologies. Dr. Samar is a former vice chair of M2M Work Group at Telecommunication Standards Development Society, India (TSDSI), an Indian SDO. He has published in various international journals, as well as conference papers, and has filed several patents. He has been invited as a guest speaker at IIT Guwahati and Defence Research and Development Organization (DRDO) India, among others. He has also been a keynote speaker at various IEEE international conferences. His current areas of research interest include software-defined networking (SDN), information-centric networking (ICN), the Internet of Things (IoT), robotics communication, fog computing, offloading, and multipath communication.

1

CENTRIC-BASED NETWORKING SYSTEMS

M. BALA KRISHNA

Contents

1.1 Centric-Based Networking Systems

1.1.1 Introduction

Centric-based networking systems (CNS) integrate the services of wired and wireless communication networks to provide secure and reliable services to end users. CNS operates in centralized and distributed modes to enhance the content delivery rate and reduce the operational and economic cost of the system. Centric systems support efficient data delivery services based on resource scheduling and service types. CNS classification is based on parameters such as user preference, information type, content type (centralized or distributed), address range, network type, network modes (device-centric or address-centric), data factor (aggregation or dissemination), and service types. CNS components and their features are explained as follows.

1.1.2 Classification of Centric-Based Networking

User-centric networking (UCN) builds an accessible network structure [1] with pre-defined rules to achieve the trade-off between the user requirements and quality of data rendered by active devices. The framework applies the content negotiation and iterative ranking techniques. The primary aspect of UCN improves the content quality value in terms of user scores (satisfaction levels). Information-centric networking (ICN) aims [2] at instantaneous control, processing, and correlation of large-scale information from the authenticated IoT devices. ICN systems ensure data availability and accessibility with minimum contention using different levels of confidentiality, security, and privacy in the system.

Current networks are confined to create end-to-end connectivity and restrict the network architecture to adapt to new service requirements and functionality. Service-centric networking (SrCN) [3] defines the "service" as the semantic information to handle the data representation, transmission, and storage in large-scale networks. A service-centric approach addresses the practical issues related to channel contention, caching, and payload transcoding. The services are categorized based on information type, query size, device type, user attributes, and application specification.

Content-centric networking (CCN) [4] supports the device mobility, maps the named-base and networking topologies, utilizes the home domain routers as proxies, and verifies the sources of data contents. CCN receivers seamlessly resend the interested data packets after the handover process and avoid the need for location updates, and the proxies update the users with the link to web pages.

Network-centric networking (NCN) [5] is a combination of wired and wireless devices that emphasize the role of handoff and location update in the network. NCN defines the collaborative and logical connections between (i) the physical and virtual network groups, (ii) service and application overlays, and (iii) implement wide-range of use cases to support the high-speed data, voice, multimedia, and internet services. NCN implements fixed topology based on the demands of access points (APs) and base stations (BSs). Network coordination, resource management, and link management aim to balance the load at different servers and BSs.

Storage-centric networking (SCN) [6] enhances the abstraction of distributed storage units located within the coverage area. In large-scale sensor networks, the data storage and retrieval operations performed at local (sensor level) and external storage systems (server level) depend on the node queries. Cluster head nodes (CHs) and the BS coordinate with local storage systems, and the data archiving systems apply lossless local data aggregation techniques to enable the fast and uniform data flow in the system. SCN is a step-wise approach that securely disseminates voluminous and multi-dimensional data to distributed storage units and provides access to authentic users.

In address-centric networking [7], the physical or IP addresses between the source and destination nodes with shortest route path indicate the end-to-end connectivity. Since the intermediate routers do not perform data aggregation task, the network traffic, latency, and resource consumption increase. However, the device-centric systems [8] modify the existing data contents (as a version) based on the requirements of the target device. A device-centric network (DCN) is based on link stability of data intensive nodes and smart devices with limited storage. DCN systems are enabled with location-based middleware services to track the devices. After the registration, the mobile devices reactively pull the data query from the server or the server proactively pushes the data updates to the mobile device. Configured routers, APs, and intermediate nodes enhance the connectivity between the data-centric nodes in the network. Figure 1.1 illustrates the classification of the centric-based networking system. This chapter elucidates the functional aspects of each centric system in the network.

1.1.3 User-Centric Networking

User-centric networking (UCN) applies pre-defined rules to improve data accessibility per user requirements [1] and defines the trade-off between user requirements and data rates. User-centric nodes are selected based on the node location, connectivity factor, buffer size, and network processing aspects. The quality and quantization steps

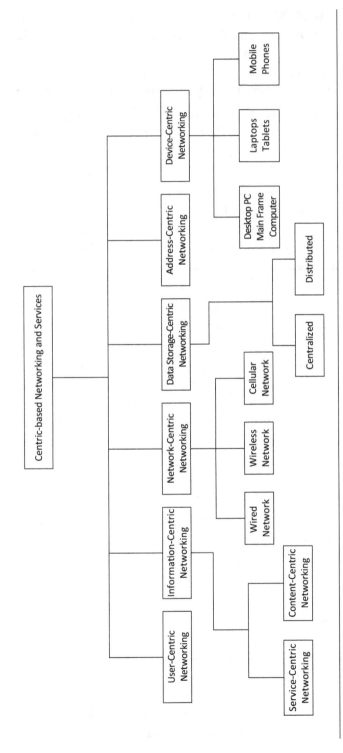

Figure 1.1   Classification of centric-based networking systems.

are evaluated to select the nodes yielding high scores. The adaptive user-centric model coordinates with content negotiation, content realization, and proxy servers to establish the decision logic to select the significant content versions and manage the user preferences. A UCN-aware throughput-based pricing scheme [9] modifies the decision rules based on content size and quality acceptable to the device type. The content delivery architecture with pre-cache feature supports multimedia services and utilizes the maximum channel bandwidth. UCN nodes [10] consider neighborhood density and location-based services to identify the active nodes in the network. The nodes are periodically verified and authenticated based on previous successful transactions. This approach eliminates malicious attacks in the network. Host nodes update the UCN status based on global information retrieved from the adjoining networks. The UCN-aware push [11] technique schedules the content delivery using unicast and multicast features of users, and the pre-cached multimedia services efficiently use the channel bandwidth. UCN identity management [12] is based on audit authority and access rules. The basic features of user-centric systems are given as follows:

- Content Adaptation: Device-centric and autonomic feature converts the existing data as per the specifications of target mobile device.
- Application Specific: UCN uses the application-specific and cooperative attributes of neighboring centric-based systems.
- Quality of Service: QoS and perceptual preferences are the primary attributes of UCN.
- Cost-effective: UCN cost-effective systems are based on node co-operations in real-time applications.

Social network analysis [13] using the UCN approach addresses the issues related to user interests, improves the data dissemination quality, and reduces the network delays. Preference-based analysis [14] with service configurations maximizes the user satisfaction levels, and infrastructure network resources efficiently handle the application workload. UCN considers the small cell BS as the dynamic group [15] that uses the cache-aided cooperative transmission, and evaluates the constraints related to cache coverage, delay, and throughput. Content popularity is based on content probability distribution at respective channels. For the centralized emergency systems, the user authentication [16] and data retrieval are reduced by considering the identity-based signature.

1.1.4 Information-Centric Networking

Machine processing and information generating systems create the interoperable cluster of machines and humans known as information of networks. ICN [17] smoothens the large information flow based on device connectivity and interaction with servers. ICN provides the persistent and location independent services to the named data objects. The network based on information, efficient resource management, and application services facilitates user queries. Information naming avoids the host-centric approach [18] that requires the users to specify the contents and server lists. ICN considers the content-centric bandwidth allocation and content delivery add-ons in the system. The traffic control autonomous systems use publish/subscribe mechanisms to address the stranded queries. The internet security protocols and filters increase the information retrieval rate and communication in large-scale networks.

Cache characterization and allocation techniques [19] compare the functionalities of optimal and opportunistic caching. The impact of the new user request is monitored and leads to subsequent requests from adjacent nodes, and the object reference for space and time is stored. This method evaluates the future temporal locality of user requests. Congestion aware caching in ICN [20] enables the node to identify the traffic flow and save the active downloads. The request popularity is represented as the exponential weighted average. When the router restarts, the previous counters of popular requests are set to zero, and the cache-based policies are updated as per the network-centric parameters. This model decouples the local searching and target searching of user requests.

ICN uses the content names [21] for smart home environments functioning as location independent in energy-efficient multi-cast system. This system supports one-one, one-to-many, many-to-one, and many-many sources to consumer communications using the extendable global namespaces with the root node represented as the logical component of unique identifier. ICN (i) ignores the identity of data origin and targets the data content; (ii) reduces the load, retrieval delays, and data contents cached from the source nodes; and (iii) addresses the services required by the IoT users. Name-based routing and lookup-based resolution fastens the content discovery mechanisms in the receiver-driven data retrieval model.

Device naming and classification in ICN [22] facilitates the human intervention to control and schedule the information management rules. This technique simplifies the actions of sensors and actuators. ICNs avoid direct connection with the end devices and support opportunistic data forwarding using multiple paths with similar contexts. This approach efficiently handles the connection inconsistencies and decouples the content source, publisher, and user nodes. ICN uses immutable data objects with specific names and properties such as streams of sensor data, collective name, and a sequence number that is easily identifiable. The devices enable the pull services when the content source has abundant information that matches the interests of participants, whereas the push model enables the participants with information that matches the interests of participants.

ICN [23] controls and categorizes the user information based on the service type. The service-oriented information is shared with the global users irrespective of their locality, migration of service, and device mobility. ICN aims at network load reduction in query-response and condition-based multicast communication in large-scale IoT networks. The security layer supports the private and confidential information in the network. ICN efficiency is based on data aggregation and data fusion that collects the data and forwards it to the data storage node or service node.

Logical virtualizations of a service-oriented ICN system [24] enable the service providers to share the virtual resources with mobile virtual network operators, and facilitate services to end-to-end devices. The information sharing between multiple SPs and MVNOs improves the fair resource sharing with low capital and operational expenditures. The resource allocation supports flexible data rates with associated users utilizing the shared bandwidth. The association indicator derived from the percentage of radio resource is shared with the active users. The cache model rewards the alleviated bandwidths utilized by the nodes, and addresses the content request rate and average data rate. This technique improves the in-network caching and energy consumption rate of active nodes in the network. In ICN-based D2D, the content caching is significant [25] since the devices depend on BS, SPs, and MVNOs. The virtual network avoids the CSI exchange between the spectral resources and backhaul network. The virtual MNOs and users directly share the contents based on the distribution indicator factor that controls the bandwidth.

1.1.5 Service-Centric Networking

Service-centric platforms [3] provide end-to-end information abstraction by (i) considering the most frequently refreshed web-pages; (ii) the quantities of encoded, transmitted information; and (iii) modifications applied to the transmitted data. The object-oriented approach distinguishes [26] the user services and contents as interests and exchanged data messages. The service entities are defined at different locations as per the requests of corresponding routers. Object-oriented service entities (OSEs) consist of processing servers to identify and separate the data and functions invoked (such as requesting the web pages with specified content) by users. OSEs encapsulate the service-centric functions and the data objects with same naming schemes. The service data cached on network elements adjacent to the mobile device is based on service calls and matches with request contents. Information transmission is categorized as private information exchange, web-cache service, and the push-based multicast transfer. Multiple service objects are rendered using (i) routing header information delivered to adjacent servers and (ii) message delays that are varied based on access costs of data objects.

The decentralized, probabilistic, and Ant Colony Optimization–based model [27] reduces the necessity of a complete network map of participant nodes. The service selection is based on matching names of current user requests to avoid the session dependency of subsequent requests. The optimal decisions are based on the distributed probability of service names and bandwidth to balance the load and minimize the network congestion. The pheromone value with the highest probability is always selected to establish the "fast-paths with best matching faces." The content-centric routing services integrate the selection and name-based routing schemes by monitoring the content names of next hops and interest packets.

The service-centric approach [28] considers the delay tolerance in do-it-yourself systems and hosts the network services. The "anywhere" cacheable service and migrated contents handle the unpredictable service disruptions and link failures. The service instances are treated as self-contained objects that exist in two states as follows: (i) static objects stored in memory and (ii) instantiated mode, currently processing the user requests. The individual services initiated by publishers as requested by subscribers are executed as unikernal virtual machines. The service execution gateway components connect the clients and application servers and execute the services of clients. The forwarding node cache, enroute the user requests, and the edge gateway node (proxy) send the requests to adjoining network. Brokers use direct or hop-by-hop store and forward techniques to transmit the messages. The service caching and synchronization allow the movement of priority service requests between the edge networks using service execution gateway nodes.

Small cell massive deployment with restricted radio frequencies in limited regions results in handover issues and degradation of the performance of large macro cells in 5G networks. The SCN-based mobility management model [29] applies the cloud-based network-centric model that uses the local information and handles the handover decisions. The user equipment receives the current location updates toward the destination. The application program interfaces collect the available radio frequencies and interference rates, and forward the call accordingly. The service-centric approach enables the application server to schedule the information in the mobile network and receive the future location information to assign the calls to the next BS. The users subscribe and send their IDs to the cloud-based driving navigation system that evaluates the arrival rate based on node mobility, and notifies the neighboring BS to initiate the handover within the coverage region. The route updates obtained as a service decrease the handover latency.

The service-centric heterogeneous cloud model [30] collaborates with the internet SPs and network SPs for optimal services that:

1. Improve the bandwidth/cost performance of distributed data centers
2. Have on-demand negotiation to map the cache contents in the user neighborhood to increase the delivery rate
3. Reduce access delays by interconnecting the deployed services
4. Configure the over-the-top (OTT) priorities and integrate the content delivery and application delivery network controllers to offload the network traffic
5. Configure the additional functions that satisfy the hardware heterogeneity and lightweight virtualization as demanded by the cloud services

The functional entities include the orchestrator to identify the available services, measure the constraints, and establish the service graphs to enhance the deployment policies. The execution zones monitor and evaluate the runtime functional aspects of service instances and coordinate with the computational resources. The deployment replicas indicate the subset of available service instances as configured by the execution zones to meet the service requirements. The service resolution subsystem logically approves the optimal service paths to address the client queries in multiple domains. The locators contain the networking address and further DNS map the location of physical devices. The out-of-band entity enables the service replicas by using standard IP address–centric connections.

1.1.6 Content-Centric Networking

Content-centric networking (CCN) systems [31,32] apply unique naming of data contents as entities (integrity and validity) that are easily retrieved by users and preserve the physical location of data servers. The caching frequency and popularity of cached contents at servers affect the system performance. The fairness of content caching [31] is improved by evaluating the contents number, cache size, average overlay path-length between the source and consumer nodes, level of content diffusion, and the cache replacement policies of the network. The performance of content overlay significantly depends on cached contents with intermediate popularities, but not on the most popular data contents. The traffic control model using fair bandwidth sharing [32] considers the data packets as a query response of user interests represented by the same name. The data content name structure includes the chunk name (with sequence number or label) for representing the data packets, whereas the object name enlists the common prefixes with version numbers of data chunks related to the corresponding named objects such as audio-video files, documents, and voice call. Caching in content networking aims at offloading the data traffic flows using one-to-one, one-to-many and many-to-one communication between the content sources and the consumers. This technique achieves the trade-off between the distributed memory capacity and the per-flow bandwidths.

The growing demands of high data rates in internet services increase the operational and control overheads of application servers. The web resources apply (i) add-on features to provide the security and location updates, (ii) low convergence time to sustain the link disruptions, and (iii) multi-stage addressing techniques to update the routing table. Content-centric networking (CCN) [4] applies the named data for initiating the receiver-driven communication and emphasizing the in-network caching. CCNs do the following: (i) efficiently map the name-based physical topology, (ii) support the receiver-driven communication that allows the users only to send the interested data packets, (iii) apply publish/subscribe mechanism using "names" to identify the desired data that match the content queries, (iv) broadcast the matching name prefixes to content routers, update the forwarding information base (FIB) and pending interest table (PIT), and (v) cache the required data to content store (CS) and coordinate with intermediate nodes to forward data toward the consumer nodes. The PIT tracks the forwarded data until the matched contents arrive at the consumer successfully, and the newly arriving contents at PIT entries replace the unmatched and timeout contents. CCN supports device mobility management by coordinating with the home domain routers that function as the proxy node and reduce the resource consumption with new domain routers. The content routers in the new domain receive the broadcast of name prefixes that are handled by the mobile source node of the home domain. This defines the prefix updates of route paths that are tunneled between the mobile content source (MCS) in home domain and content routers (CRs) of the new domain.

The content-centric approach [33] explores the content delivery network for high data rates, and filters the redundant information in the network. The performance of a content delivery network depends on the coordination of the basic delivery network model, content providers, and network operators. A content delivery network does the following: (i) deploys the servers, (ii) expands the server content list as per the user demands, (iii) collaborates with multiple internet SPs and network operators, and (iv) reduces the communication gaps between different content delivery network systems.

CCN [34] enables the routers to retransmit the dropped and pending data packets periodically. The lifetime of interest lists in the pending table is enhanced by the entry time for keeping the upstream alive, and until the request data is received. The pending interests are periodically retransmitted to consider the timeout requests of consumers. Content objects [35] match the identical data contents (frequently repeated words, phrases, songs, or pictures) from the distributed information among the neighboring servers. The composite information is derived using limited resources. Content security [36] is achieved by using the name sequence with implicit digest and public signatures. The forwarding NACKs verify the status of the next forwarding node, and the NACKs reduce the drawbacks of flooding. The progressive role of content sources in internet systems requires the autonomous content and cache systems [37] to coordinate and improve the peer business relations. The global performance of content-centric networks is achieved through multi-dimensional distributed cache systems based on (i) cache decision policies that coordinate with resources located at different places, (ii) the cache allocation matching the user requests, reducing the pending interests, and (iii) global stabilization of cache configurations. Content peering among the autonomous internet SPs exchange the cache summary, locally available messages, and share the most recent interest copies. The cache savings and cost minimization depend on the accuracy of the arrival time request. Cache synchronization is based on timely updates and achieves the uniform allocation of cache resources.

The crowd source–based CCN [38] meets the demands of mobile participant nodes by balancing the human intelligence and server capacity. The social networking characteristics such as content sharing with community groups aggregate the interesting contents and readily diffuse the information to the sensor network. The mobile devices registered at the application entity (access points) act as content sources to participate in the task. The participants with similar interests join the task and forward the shared data contents. The task initiator node evaluates the user participation in the group, and the feedback is sent to the cloud server. The user equipment communicates with the service initiation using cellular and social group communication.

1.1.7 Network-Centric Networking

Network-centric networking (NCNs) [5] decentralizes the sparse and structured wireless mesh networks using real and virtual resources as overlay services and establishes the end-to-end connectivity. NCN is a combination of wired and wireless networks that support high-speed connectivity in multimedia and internet services. Network-centric systems evolved from centralized to distributed mode include the broad participation of wireless sensors, actuators, personal computers, mobile phones, smart objects, RFIDs, and satellite systems. SPs and network operators control the network deployment, operational strategies, link management, and service pricing schemes. SPs separate the core capital network services with access and edge networking services operated from different locations.

The network-centric network aims for optimal coverage [39] to balance the load and resource allocation at each radio access point. Deployment strategies include the decision logic models such as Semi-Markov Decision and joint radio resource management schemes that evaluate the network performance in terms of resource blocking cost, network throughput rate, and financial gain. The deployment model considers the (i) logical partition of cellular regions into finite concentric rings, (ii) enable the mobile users to use identical radio conditions for signal-to-noise ratio evaluation (SNR measurements based on path loss and Rayleigh fading), (iii) placement of BS in overlapping regions of two radio access points (using WiMAX and UMTS technologies), (iv) effects of intra-cellular interference and inter-cellular interference, and (v) evaluation of the orthogonality and path loss of users in multiple adjacent rings to improve the traceability of mobile users. The cell blocking cost depends on call arrival rates, state transition probability, and the sojourn times (expected time for each user to associate with the corresponding cell). The calls are directed toward the cell that increases the network revenue and directly impacts the rewards offered to respective users.

The network-centric collaborations indicate the similarity of information, trust relations, and interaction frequency of social group members. PopCore, the network-centric collaborative platform [40], uses Facebook data to derive useful information related to favorite items. The duplicated and uncategorized web pages related to individuals are considered to understand the behavior of social group members. The network algorithms are applied to (i) randomly select the liked itemsets from a friend's profile of users, (ii) select the most liked items, (iii) select the most liked items of social groups members, (iv) provide most-recent to recent-past (current last week) user information, (v) yield similarity scores for the most liked and disliked items, and (vi) measure the interaction strength that indicates the strong association of a user's views for selected items.

The network-centric situational awareness system [41] aims to optimize network performance and improve the quality of experience (QoE) for end users that are further evaluated as subjective feedback of services such as data speed as well as the quality of audio and video files. The surrounding, target, location, and responsiveness model addresses the end-to-end video communication in wireless ad hoc networks. Human factors such as observing, understanding, and recalling event levels define the relevance and user responsiveness, and further update the network. The surrounding awareness and location indicate user responsiveness to environmental changes. The information fusion depends on the rate of packet loss and signal delays of streaming videos in the physical communication channel. NCNs with situational awareness define the object functions that address the resource allocation issues as required by the end users in emergency situations.

The characteristics of NCNs are categorized as follows:

- Network-centric systems provide a unique IP address for the connected devices such as sensors, actuators, and RFIDs to select the shortest route between the communicating entities.

- NCN classifies the user's performance based on sensing the environment and defines the autonomous actions specific to a user. The traffic flow and scheduling are varied as per the social emergence and network requirements.
- The decentralized networking allows the sensor nodes to actively control the services, vary the packet delivery rate, and coordinate with the BSs to manage the resources.
- In social networking, the collaborative filters provide useful recommendations to choose the online products; movie, audio/video albums, books, and so on.

1.1.8 Storage-Centric Networking

Storage-centric networks are based on location and the storage type used in the network. Data storage systems manage the flow of data in centralized, distributed, and zone-based modes to support multiple user queries in the large-scale network. SCN uses the core router, access router, and end-point routers to implement the tree-based topology and recursive-based topology in the network. Sensor storage capacities depend on the rechargeable batteries embedded in the sensor system, and the storage capacity of sensing devices [6] depends on battery energy levels. Adaptive designs enable the sensor nodes to proactively sense, perform local actions, and store the current events. The secure sensor nodes buffer the events of physical environment based on the user query and naming services of the network. Storage systems use the current database server as the resource monitor to identify the (i) active neighbors, (ii) existence of reliable route paths between the source and destinations nodes, and (iii) number of duplicate messages currently present at the sensor nodes. This ensures the availability of current information at the local data center. The message frequency is varied based on the ratio of incoming user queries, with the regular updates stored at respective local databases. BS depends on the ratio of received replies, selects the number of active storage nodes, and periodically broadcasts the sensor messages. The data chunks received within the time limit are verified with the node and stream IDs. The valid data chunks are sent to local files that are stored in the virtual directory of BS. The nodes track the successfully sent data chunks at these local files and delete the current data from the storage.

Stream-oriented data storage systems [42] decouple the data processing from storing the data. The adaptable application-specific policies are used for distinct storage areas in the system. The sensor nodes modify the sensing time intervals and frequency to maintain the current energy at threshold levels and update the BS. This technique enables the sensor nodes to (i) sustain the bandwidth limitations, (ii) be active in the sense-aggregate-send loop, and (iii) retain the data until it is transmitted to the BS. In large-scale sensor networks, the data updates are flooded to the storage systems since the sensor nodes dynamically vary their active and sleep schedules. This requires the data dissemination [43] and topological variance between the tree-based and grid-based structures. In the network, the sensing zone and storage zone [44] reduce message replications and ensure reliable data rates with the quality of information at the storage centers. The storage manager nodes use the linear coefficient of data packets to yield the encoding vectors and distribute the data packets.

1.1.9 Address-Centric Networking

Address-centric [7] networks require high location accuracy and facilitate the end-to-end connectivity using the shortest route path between the source and the destination nodes. This method does not consider the data aggregation to be performed at the sensor level and directly transmits the data to sink nodes. This creates data redundancy that is further avoided by the data-centric approach. The source sends identification information to the sink node, whereas in a data-centric approach the source sends distinct information to the sink node. IP subnet addressing modes [45] increase the deployment cost of the sensor network. The subnet nodes coordinate with the home and foreign access routers, server nodes, and end-user nodes. In static networks, the routers periodically update the routing tables, whereas in Mobile IP, the Care of Addresses (CoAs) is updated based on the foreign network of the mobile device. Increases in IoT sensors [46] and 5G networks with a unique addressing scheme escalate the complexity of channel distribution. The complexity in the address-centric network is due to the locator module using multicast routing addresses. Openflow switches in software-defined networks [47] use IP addressing schemes to update the flow tables with IP source and port address. This approach considerably increases the flow at switches, but doesn't monitor the redundant data at switches.

1.1.10 Device-Centric Networking

The network-centric approach coordinates between the gateway and router nodes. Device-centric systems enable the mobile network operators [8] to configure the resources as per user demands. While coordinating with the component-based system, a device-centric cloud-based platform facilitates the content services [48] to mobile devices. This technique supports the reusable components in network architecture. Service intra-organization of connected devices secures the ecosystem. Device-centric architecture [49] supports the database deployment using multiple transmit signals in the cellular network. The control plane using high power nodes as microwave transmissions is separated from the data plane. Low-power nodes in the data plane consider millimeter wave transmissions.

High data rates and increasing traffic volumes in 5G networks require low latency and application-specific deployment (with user equipment connected to at least one BS within the coverage area). Due to random mobility, the links are sustained by a cloud-based device-centric system [50] using radio access virtualization. This technique continuously tracks the intensity and beam width of the user equipment. User plane virtualization for downlink is supported by the hyper-transceiver system to improve the quality of service. This model supports multipoint transmission and proactive dynamic point selection to connect the devices in short-range communication.

In large-scale heterogeneous networks, the operational and eco-nomical costs reduce the cellular offloading by lowering the device-to-device radio links [51]. The connection establishment with faster delivery rate archives (i) call offloading, (ii) sharing of game and multi-media features, (iii) context-aware services, and (iv) security alerts and notifications in social groups. DCN uses the network knowledge and supports the distributed traffic in peak hours [52] to enable the user equipment to utilize the services based on previous sessions. A device-centric system with link management decisions uses the coordinated multiple heterogeneous devices to configure the femto APs [53] in a single-tier cellular network with multiple BS. A device-centric system with secure deployment and operations [54] supports the device-aware constraints such as risk context identification to secure the system.

1.1.11 Conclusion and Future Directions

Centric systems consider the attributes of user, data, information, storage, service, and network with device-specific and address-specific features in communication network. This chapter classifies the centric systems and highlights the features and functionality of each centric system. User-centric systems consider the trade-offs between user preferences and quality of information received from the network. The information-centric system aims at data availability based on contexts, and is further classified as service-centric or content-centric network. Network-centric networks define the collaborative topology as per the policies of SPs and network operators. The network deployment is based on rules specified by the business model. The link management and traffic scheduling schemes are adequately controlled and coordinated by the BSs and APs. Storage-centric networks aim at reducing the complexity of data processing, storing the data with minimum replica and reducing the data loss due to disruptive communication. The address-centric networks use IPv4 and IPv6 addresses to define the shortest route paths. Device-centric networks enable deployment, resource management, and link management as per user demands. Future extension of this chapter includes the features of fog computing, edge computing, and multi-access edge computing in centric-based networking systems.

References

1. Lum W. Y. and Lau F. C. M., "User-Centric Content Negotiation for Effective Adaptation Service in Mobile Computing," *IEEE Transactions on Software Engineering*, Vol. 29(12), December 2003, pp. 1100–1111.
2. Rayes A., Morrow M., and Lake D., "Internet of Things Implications on ICN," In *Proceedings of IEEE International Conference on Collaboration Technologies and Systems (CTS)*, Denver, CO, 21–25 May 2012, pp. 27–33.
3. Wolf T., "Service-Centric End-to-End Abstractions in Next-Generation Networks," In *Proceedings of IEEE 15th International Conference on Computer Communications and Network (ICCCN)*, Arlington, VA, 9–11 October 2006, pp. 79–86.

4. Lee J., Cho S., and Kim D., "Device Mobility Management in Content-Centric Networking," *IEEE Communications Magazine*, Vol. 50(12), December 2012, pp. 28–34.

5. Vicente J., Rungta S., Ding G., Krishnaswamy D., Chan W., and Miao K., "OverMesh: Network-Centric Computing," *IEEE Communications Magazine*, Vol. 45(2), February 2007, pp. 126–133.

6. Wang L., Noh D. K., Yang Y., Le H. K., Abdelzaher T. F., and Ward M., "AdaptSens: An Adaptive Data Collection and Storage Service for Solar-Powered Sensor networks," In *Proceedings of IEEE 30th Real-Time Systems Symposium (RTSS)*, Washington, DC, 1–4 December 2009, pp. 303–312.

7. Son D., Helmy A., and Krishnamachari B., "The Effect of Mobility-induced Location Errors on Geographic Routing in Mobile Ad Hoc and Sensor Networks: Analysis and Improvement Using Mobility Prediction," *IEEE Transactions on Mobile Computing*, Vol. 3(3), July–August 2004, pp. 233–245.

8. Kupper A., Treu G., and Linnhoff-Popien C., "TraX: A Device-Centric Middleware Framework for Location-Based Services," *IEEE Communications Magazine*, Vol. 44(9), September 2006, pp. 114–120.

9. Feng N., Mau S.-C., and Mandayam N. B., "Pricing and Power Control for Joint Network-Centric and User-Centric Radio Resource Management," *IEEE Transactions on Communications*, Vol. 52(9), September 2004, pp. 1547–1557.

10. Boettner P., Gupta M., Wu Y., and Allen A. A., "Towards Policy Driven Self Configuration of User Centric Communication," In *Proceedings of ACM 47th Annual Southeast Regional Conference*, Clemson, SC, 19–21 March 2009, pp. 35:1–35:6.

11. Bhatia R., Narlikar G., Rimac I., and Beck A., "UNAP: User-Centric Network-Aware Push for Mobile Content Delivery," In *Proceedings of IEEE 28th Conference on Computer Communications (INFOCOM)*, Rio de Janeiro, Brazil, 19–25 April 2009, pp. 2034–2042.

12. Vossaert J., Lapon J., Decker B. D., and Naessens V., "User-Centric Identity Management Using Trusted Modules," *Elsevier Journal of Mathematical and Computer Modelling*, Vol. 57(7–8), April 2013, pp. 1592–1605.

13. Gao W. and Cao G., "User-Centric Data Dissemination in Disruption Tolerant Networks," In *Proceedings of IEEE 30th Conference on Computer Communications (INFOCOM)*, Shanghai, China, 10–15 April 2011, pp. 3119–3127.

14. Garcia-Galan J., Pasquale L., Trinidad P., and Ruiz-Cortes A., "User-Centric Adaptation Analysis of Multi-tenant Services," *ACM Transactions on Autonomous and Adaptive Systems (TAAS)—Special Section on Best Papers from SEAMS*, Vol. 10(4), February 2016, pp. 24:1–24:25.

15. Zhang H., Chen Y., and Yang Z., "Hierarchical Cache-Aided Transmission Cooperation in 5G User-Centric Network: Performance Analysis and Design Insights," *Elsevier Journal of Network and Computer Applications*, Vol. 111, 1 June 2018, pp. 17–27.

16. Li R., Asaeda H., Li J., and Fu X., "A Distributed Authentication and Authorization Scheme for In-Network Big Data Sharing," *Elsevier Journal of Digital Communications and Networks*, Vol. 3(4), November 2017, pp. 226–235.

17. Dannewitz C., Kutscher D., Ohlman B., Farrell S., Ahlgren B., and Karl H., "Network of Information (NetInf)—An Information-Centric Networking Architecture," *Elsevier Journal of Computer Communications*, Vol. 36(7), April 2013, pp. 721–735.

18. Xylomenos G., Ververidis C. N., Siris V. A., Fotiou N., Tsilopoulos C., Vasilakos X., Katsaros K. V., and Polyzos G. C., "A Survey of Information-Centric Networking Research," *IEEE Communications Surveys & Tutorials*, Vol. 16(2), Second Quarter 2014, pp. 1024–1049.

19. Dabirmoghaddam A., Mirzazad-Barijough M., and Garcia-Luna-Aceves J. J., "Understanding Optimal Caching and Opportunistic Caching at 'The Edge' of Information-Centric Networks," In *Proceedings of ACM 1st Conference on Information-Centric Networking (ACM-ICN)*, Paris, France, 24–26 September 2014, pp. 47–56.

20. Badov M., Seetharam A., Kurose J., Firoiu V., and Nanda S., "Congestion-Aware Caching and Search in Information-Centric Networks," In *Proceedings of ACM 1st Conference on Information-Centric Networking (ACM-ICN)*, Paris, France, 24–26 September 2014, pp. 37–46.

21. Amadeo M., Campolo C., Iera A., and Molinaro A., "Information Centric Networking in IoT Scenarios: The Case of a Smart Home," In *Proceedings of IEEE International Conference on Communications (ICC), SAC-Internet of Things*, London, UK, 8–12 June 2015, pp. 648–653.

22. Lindgren A., Abdesslem F. B., Ahlgren B., Schelen O., and Malik A. M., "Design Choices for the IoT in Information-Centric Networks," In *Proceedings of IEEE 13th Annual Consumer Communications & Networking Conference (CCNC)*, Las Vegas, NV, 9–12 January 2016, pp. 882–888.

23. Chen J., Li S., Yu H., Zhang Y., Raychaudhuri D., Ravindran R., Gao H., Dong L., Wang G., and Liu H., "Exploiting ICN for Realizing Service-Oriented Communication in IoT," *IEEE Communications Magazine*, Vol. 54(12), December 2016, pp. 24–30.

24. Liang C. and Yu F. R., "Virtual Resource Allocation in Information-Centric Wireless Virtual Networks," In *Proceedings of IEEE International Conference on Communications Mobile and Wireless Networking Symposium*, London, UK, 8–12 June 2015, pp. 3915–3920.

25. Wang K., Yu F. R., and Li H., "Information-Centric Virtualized Cellular Networks with Device-to-Device (D2D) Communications," *IEEE Transactions on Vehicular Technology*, Vol. 65(11), November 2016, pp. 9319–9329.

26. Braun T., Hilt V., Hofmann M., Rimac I., Steiner M., and Varvello M., "Service-Centric Networking," In *Proceedings of IEEE International Conference on Communications Workshops (ICC)*, Kyoto, Japan, 5–9 June 2011, pp. 1–6.

27. Shanbhag S., Schwan N., Rimac I., and Varvello M., "SoCCeR: Services over Content-Centric Routing," In *Proceedings of ACM SIGCOMM workshop on Information-centric networking*, Toronto, ON, 19 August 2011, pp. 62–67.
28. Sathiaseelan A., Wang L., Aucinas A., Tyson G., and Crowcroft J., "SCANDEX: Service Centric Networking for Challenged Decentralised Networks," In *Proceedings of ACM Workshop on Do-it-yourself Networking: An Interdisciplinary Approach (DIYNetworking), The 13th International Conference on Mobile Systems, Applications, and Services (MobiSys)*, Florence, Italy, 18 May 2015, pp. 15–20.
29. Sueda Y. and Koike A., "Service Centric Mobility Management for Improving Quality of Experience toward Future Mobile Network," In *Proceedings of IEEE International Symposium on Local and Metropolitan Area Networks (LANMAN)*, Rome, Italy, 13–15 June 2016, pp. 1–6.
30. Simoens P., Griffin D., Maini E., Phan T. K., Rio M., Vermoesen L., Vandeputte F., Schamel F., and Bursztynowski D., "Service-Centric Networking for Distributed Heterogeneous Clouds," *IEEE Communications Magazine*, Vol. 55(7), May 2017, pp. 208–215.
31. Tortelli M., Cianci I., Grieco L. A., Boggia G., and Camarda P., "A Fairness Analysis of Content Centric Networks," In *Proceedings of International Conference on the Network of the Future*, Paris, France, 28–30 November 2011, pp. 117–121.
32. Oueslati S., Roberts J., and Sbihi N., "Flow-Aware Traffic Control for a Content-Centric Network," In *Proceedings of IEEE INFOCOM*, Orlando, FL, 25–30 March 2012, pp. 2417–2425.
33. Carofiglio G., Morabito G., Muscariello L., Solis I., and Varvello M., "From Content Delivery Today to Information Centric Networking," *Elsevier Journal of Computer Networks*, Vol. 57(16), 13 November 2013, pp. 3116–3127.
34. Abu A. J., Bensaou B., and Wang J. M., "Interest Packets Retransmission in Lossy CCN Networks and its Impact on Network Performance," In *Proceedings of ACM First Conference on Information-Centric Networking*, Paris, France, 24–26 September 2014, pp. 167–176.
35. Liu H., Chen Z., Tian X., Wang X., and Tao M., "On Content-Centric Wireless Delivery Networks," *IEEE Wireless Communications Magazine*, Vol. 21(6), December 2014, pp. 118–125.
36. Compagno A., Conti M., Ghali C., and Tsudik G., "To NACK or not to NACK? Negative Acknowledgments in Information-Centric Networking," In *Proceedings of IEEE 24th International Conference on Computer Communication and Networks (ICCCN)*, Las Vegas, NV, 3–6 August 2015, pp. 1–10.
37. Pacifici V. and Dan G., "Coordinated Selfish Distributed Caching for Peering Content-Centric Networks," *IEEE/ACM Transactions on Networking*, Vol. 24(6), December 2016, pp. 3690–3701.

38. Wang K., Gu L., Guo S., Chen H., Leung V. C. M., and Sun Y., "Crowdsourcing-Based Content-Centric Network: A Social Perspective," *IEEE Network Magazine,* Vol. 31(5), 28 September 2017, pp. 28–34.
39. Ibrahim M., Khawam K., and Tohme S., "Network-Centric Joint Radio Resource Policy in Heterogeneous WiMAX-UMTS Networks for Streaming and Elastic traffic," In *Proceedings of IEEE Wireless Communications and Networking Conference (WCNC),* Budapest, Hungary, 5–8 April 2009, pp. 1–6.
40. Sharma A. and Cosley D., "Network-Centric Recommendation: Personalization with and in Social Networks," In *Proceedings of IEEE Third International Conference on Privacy, Security, Risk and Trust (PASSAT) and IEEE Third International Conference on Social Computing,* Boston, MA, 9–11 October 2011, pp. 282–289.
41. Saadou A. and Chenji H., "A Network-Centric Model of Situational Awareness," In *Proceedings of IEEE Military Communications Conference (MILCOM),* Baltimore, MD, 23–25 October 2017, pp. 73–78.
42. Mottola L., "Programming Storage-Centric Sensor Networks with Squirrel," In *Proceedings of ACM/IEEE 9th International Conference on Information Processing in Sensor Networks (IPSN),* Stockholm, Sweden, 12–16 April 2010, pp. 1–12.
43. Luo H., Xing G., Li M., and Jia X., "Dynamic Multi-resolution Data Dissemination in Storage-Centric Wireless Sensor Networks," In *Proceedings of ACM 10th Symposium on Modeling, Analysis, and Simulation of Wireless and Mobile Systems (MSWiM),* Chania, Crete Island, 22–26 October 2007, pp. 78–85.
44. Nguyen N., Krishnamurthy S., Xie P., and Jones D., "Improving Information Availability in Storage-Centric Sensor Networks," In *Proceedings of IEEE 33rd Conference on Local Computer Networks (LCN),* Montreal, Que, October 14–17, 2008, pp. 82–90.
45. Tubaishat M. and Madria S. K., "Sensor Networks: An Overview," *IEEE Potentials,* Vol. 22(2), April-May 2003, pp. 20–23.
46. Amadeo M., Campolo C., Quevedo J., Corujo D., Molinaro A., Iera A., Aguiar R. L., and Vasilakos A. V., "Information-Centric Networking for the Internet of Things: Challenges and Opportunities," *IEEE Network Magazine,* Vol. 30(2), March–April 2016, pp. 92–100.
47. Luo T., Tan H.-P., and Quek T. Q. S., "Sensor OpenFlow: Enabling Software-Defined Wireless Sensor Networks," *IEEE Communications Letters,* Vol. 16(11), November 2012, pp. 1896–1899.
48. Raatikainen M., Komssi M., Bianco V. D., Kindstom K., and Jarvinen J., "Industrial Experiences of Organizing a Hackathon to Assess a Device-Centric Cloud Ecosystem," In *Proceedings of IEEE 37th Annual Computer Software and Applications Conference (COMPSAC),* Kyoto, Japan, 22–26 July 2013, pp. 790–799.
49. Boccardi F., Heath R. W., Lozano A., Marzetta T. L., and Popovski P., "Five Disruptive Technology Directions for 5G," *IEEE Communications Magazine,* Vol. 52(2), February 2014, pp. 74–80.

50. Maaref A., Ma J., Salem M., Baligh H., and Zarifi K., "Device-Centric Radio Access Virtualization for 5G Networks," In *Proceedings of IEEE Globecom Workshops (GC Wkshps): Emerging Technologies for 5G Wireless Cellular Networks*, Austin, TX, 8–12 December 2014, pp. 887–893.
51. Andreev S., Pyattaev A., Johnsson K., Galinina O., and Koucheryavy Y., "Cellular Traffic Offloading onto Network-Assisted Device-to-Device Connections," *IEEE Communications Magazine*, Vol. 52(4), April 2014, pp. 20–31.
52. Jiang M., "Device-Controlled Traffic Steering in Mobile Networks," In *Proceedings of IEEE 9th International Conference on Next Generation Mobile Applications, Services and Technologies*, Cambridge, UK, 9–11 September 2015, pp. 7–12.
53. Yeh S.-P., Talwar S., Wu G., Himayat N., and Johnsson K., "Capacity and Coverage Enhancement in Heterogeneous Networks," *IEEE Wireless Communications Magazine*, Vol. 18(3), 16 June 2011, pp. 32–38.
54. Chen C. and Helal S., "A Device-Centric Approach to a Safer Internet of Things," In *Proceedings of ACM International workshop on Networking and object memories for the internet of things (NoME-IoT)*, Beijing, China, 18 September 2011, pp. 1–6.

2

NAMING AND ADDRESSING IN INFORMATION-CENTRIC NETWORKING

G. AKHIL AND SAMAR SHAILENDRA

Contents

2.1 Introduction

In today's world, the role of the Internet cannot be ignored. This spans across the communication between any two systems in the form of bits over different kinds of networks across the world both wired and wireless in nature. The Internet is considered to run over a legacy IP Protocol. A brief recap as to what this constitutes is: The Internet follows a basic four-layer architecture. The layers are the application layer (consisting of both the presentation and the session layer if one is studying the seven-layer Open System Interconnect (OSI) model), the transport layer, the Network or Internet layer, and the Network Interface layer (jointly represents the data link layer and physical layer of OSI model). The responsibilities of each layer are: The application layer hosts the user applications. These applications may use protocols like ftp, http, etc. for exchanging data with the remote host. This layer uses various services of Transport layer for reliable and efficient data transport. The two most popularly used protocols at the Transport layer are Transmission Control Protocol (TCP) and User Datagram Protocol (UDP). The former is known for its reliability (provides end-to-end error recovery using retransmission) while the later provides smaller delays at the cost of reliability. This layer logically connects the applications at two ends. The next layer in the Internet layer architecture is the Network/Internet layer. The primary function of this layer is to route the traffic through the network. The layer after this is the Network Interface Layer. This layer does the low-level functionalities of Medium Access Control at the link level, error recovery of any lost frames, and the actual physical transmission of the data on the underlying link. Unlike above layers, the scope of this layer is only till next hop and provides hop by hop control over the medium and frame error if any. It is worthwhile to note that users' requirements in today's world have changed significantly from those of the time when the current Internet architecture was conceived. Though by far the current architecture is able to satisfy all our requirements, it won't be long until it will be not able to meet the exponentially growing demands of modern users. The number of systems around the world is increasing greatly but apart from this, the number of other devices that connect to the Internet is also increasing greatly. Laptops, mobile devices, cloud-based services, IoT, etc. are all concepts that have increased the number of devices connecting to the Internet significantly. According to surveys, there are going to be more than 20 billion IoT devices and smartphones by 2020. This huge number cannot be handled by the current legacy architecture for a variety of reasons like namespace, traffic load, effective traffic management, delays, and a variety of other reasons.

Apart from this, even the type of traffic that is flowing through the Internet has changed significantly over time. There is huge demand for video-based traffic as opposed to the traditional text-based web page traffic, file transfers, and e-mail. With the advent of cloud services, online music streaming, gaming, content sharing, storage, etc. are all becoming interesting technologies that are pushing the limits of the Internet. Apart from this, P2P clients like Bit Torrent etc. also are changing the type of traffic that is being carried over the Internet today.

In recent times, the number of users connected to the Internet also has increased significantly. Moreover, the Quality of Service (QoS) desired by today's consumer cannot possibly be handled by the current Internet architecture. The traditional network also does not give us any advantages that we can make use of apart from a hierarchical naming structure in the form of an IP address. The Internet, which has already been patched several times to accommodate the users' demands, is left with hardly any scope for further improvement related to the issues of effective packet management, latency, security, etc.

Researchers have realized these challenges and a new architecture paradigm called Information-Centric Networking (ICN) has been proposed in the literature.

The traditional TCP/IP considers the session/link between client and server as the main point of focus. It considers the role of networking to be creating and maintaining the logical link between the client and the server in which the data transfer happens. The methods to understand what data is being requested, the server from which this data should be requested, how far to communicate, etc. are all decided using additional protocols depending upon the required and supported functionalities. Thus, in this form of networking, the emphasis is placed on the server itself rather than the data that is being requested by the user.

ICN changes the point of focus and importance from the server to the data itself that is being requested. In a network that follows the concept of ICN, the data that is being requested is considered to be an object called a data object and is given a name which helps users of the Internet to identify the object. In a typical ICN scenario, the user queries for a data object, say a song, in the network in a manner similar to a Google search. The search returns a list of songs, each of which is connected to the original song requested by the user in some manner. This is because data objects uploaded by different users may have different names and thus there exist differently named data objects which are essentially the same song as the user requested. These are returned by the network to the user. The user can look at the songs that have been displayed and pick one of them as the song he wants depending on either the attributes/keywords or the identification of the name of the object. Once selected, a request is sent into the network asking for the song that the user wishes to query about and download. The network smartly searches its caches as well as the closest server(s) and provides the song requested. The difference is that today the request is sent to a single server which responds back with the song. The network is unaware of the data that is being sent or received and thus cannot work in a smart/intelligent manner to increase efficiency for the queries of this data. In contrast to this, in ICN paradigm, the network works smartly and does not necessarily get the whole song from the same server. It rather downloads it in parts from multiple replicas of the same song in parallel.

This is enabled through the use of caching. Caching is envisioned as one of the key enabler technologies for smart decision making in ICN. With the help of caching, it is possible for routers themselves to service the requests of the user by having the routers cache data objects. This reduces the need to connect to the server for each individual user thus freeing up valuable resources for more optimal usage and better service. This affects performance in a varied manner like reducing the delay experienced by the user, freeing up network resources and making service more efficient in ICN networks. For a better idea about the importance of caching, readers can read [25,26].

An example to explain the situation with and without ICN would be as follows: Assume A and B are sitting in adjacent dormitory rooms trying to query for the latest episode of a Netflix show, i.e., they are both trying to watch the same episode. In traditional networking, both would request the connection to Netflix's server and ask for a copy of the same episode individually. This would imply two connections through the same routers but two different end hosts to the same server for the same data. The intervening network does not know this so cannot optimize itself to work in a better fashion and thus allows the double usage of resources. Thus, in the end, both A and B get the required episode that they watch but each of them uses up some amount of network resources in terms of processing power, link usage, etc. This implies that the total network usage is twice that being used by A and B individually. On the other hand, in ICN paradigm, the router en-route may possibly cache the data while the first user requests the episode. It serves from the local cache to the second user upon request of the same data. This in turn significantly reduces the network resources being consumed. Moreover, the savings become manifold as the number of users requesting the same content increases.

This is demonstrated by viewing Figures 2.1 and 2.2. Figure 2.1 displays the traditional data transfer for a data object A in a legacy network when requested by three different users. As shown in Figure 2.1, all three laptops make individual requests and responses to the server.

As opposed to this, Figure 2.2 shows the same scenario where three different laptops are querying for the same object in an almost simultaneous manner like Figure 2.1. Please notice the difference in Figure 2.2 where instead of server, the query is fulfilled by the network. This is important because the ICN paradigm shifts away focus

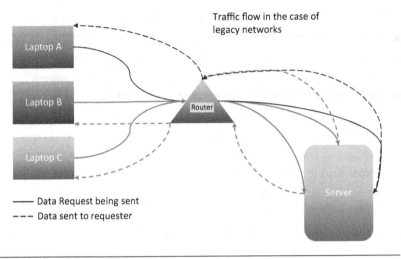

Figure 2.1   Traffic flow in legacy networks.

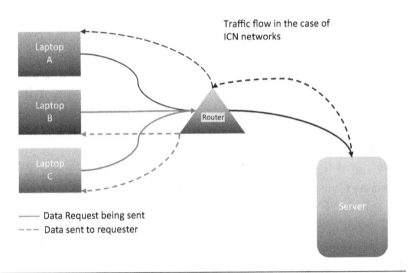

Figure 2.2   Traffic flow in ICN networks.

from single server needing to fulfill the request to clever techniques in network that can smartly handle the request for the data object requested.

Due to the increase in the number of devices there is an increase in the number of queries for data into the network. To handle this exponentially growing traffic, the network needs to effectively handle the data without causing a huge load on the network resources.

This requires smart handling of the data object that is being requested. In the current paradigm, such smart network resource management does not seem possible since the focus is not on the data, whereas unlike traditional networks, in the ICN paradigm the data is the focus; thus the network can intelligently manipulate queries through the use of efficient caching, routing, naming, etc [33].

There are many factors that need to be considered when one is concerned with naming and addressability in the information-centric networking paradigm. Covered in the following sections are some challenges and requirements of naming schemes, a few classes of naming schemes, some actual algorithms used, and questions related to security, mobility, and wireless conditions. Each of these are important understand as they relate to naming schemes and the role they play [5].

2.2 Challenges in Naming

2.2.1 Introduction to Naming

The concept of naming is an old one. A MAC address is a kind of name for a computer, an IP address is another kind of name which identifies a system on the Internet. Each of these names serves certain purposes that need to be fulfilled like identifying a system for routing, switching, etc.

An IP address is a 32 bit name given to a system to identify that system and use a hierarchical mechanism of addressing to enable routing to that system to ensure that the system gets connected to the Internet. A similar kind of function, which does not use hierarchical routing, is used by MAC for switching.

In the ICN paradigm, the focus is shifting from the server providing the data/information to the information itself. We query for a specific kind of information, we search for it, we download it, and we focus on that information. But there needs to be a mechanism for the network to identify this information. In ICN, such information is called a *data object*.

Naming is a mechanism by which we assign an identifier to the various data objects, which can enable us to perform a few functions on them. Such identifiers can be given in a variety of ways, each of which enables us to perform a few functions while make performing other functions more tedious. But before we look at the different kinds of ways of naming, we need to identify the challenges that a naming mechanism faces and the requirements that a naming mechanism ideally needs to fulfil. The challenges are listed in the following subsections.

2.2.2 Identification of Data

The first challenge is to uniquely identify the data object. There are multiple objects available and multiple versions of the same kind of data. For example, different people upload different versions of the same song. This song should now be identified by different names because it has been uploaded by different people. But the name given to the song should somehow capture the fact that the song still is the song that is being searched for and uploaded. The name should be able to uniquely identify the song uploaded by uploader A and mark it as different from the song uploaded by uploader B. This is especially difficult if one keeps in mind the large amount of data already available on the Internet and the large amount of data that is being uploaded on a daily basis. Everything from a picture in the news to a video on any video streaming service to text documents needs to be captured in a similar manner and if there are multiple copies of it uploaded by different people, each of those copies also should be considered individually.

This becomes an especially difficult task because it requires a huge namespace to be able to name that many objects. To give you an idea of how difficult it is, we are currently shifting to IPv6 because IPv4 is running out of namespace to name all the individual devices on the Internet. Now consider that each of those devices is uploading more than one piece of data and we get an idea of how much naming is required.

2.2.3 Assistance in Routing

Another challenge is that of searching and efficiently routing the data object based on its name. To effectively locate an object, we need an address in simple language. This address is an identifier that can be the name of the object, or it may not be the name of the object. A lot of debate is still happening on whether the identifier and the locator should be the same but that is outside the scope of discussion of this chapter. For the purposes of this chapter, we will assume that the identifier and locator are both the name of the object itself.

The current paradigm uses IP and servers. The IP address effectively identifies which server to connect to. The onus of providing content is always on the server and the network only facilitates the delivery of the data packets by appropriately searching the IP address based upon the routing table at the routers. In the ICN paradigm where data is given preference and the servers do not play such a major role, the onus is on the network to locate the data object itself and route it to the required location. The network now needs to be smart enough to identify the data object that is being searched for, locate the data object in the network, and be able to effectively devise a route to deliver the object to the requester. This is not an easy task and the naming scheme needs to play a role in facilitating this process [7,29].

2.2.4 Security and Privacy

Security is major concern in today's networks. Recently, there have been a lot of cyberattacks and scams. This is a growing issue with a lot of focus being given to it. From a security perspective, there are certain considerations that need to be taken when shifting to the ICN paradigm. The first is that it is imperative to identify and certify data objects as the correct objects, especially since the network is now trusted and smart and able to identify data objects. In traditional networks, the server is certified and identifiable. The same should carry over to each individual data object so that it cannot be faked and there is a guarantee that the object is what it says it is and the identity of the object is also certifiable. This facility should be somewhat aided by the name.

2.2.5 Interconnection with Architecture

A naming scheme alone does not have much of an impact if the underlying architecture of the network does not support it. This is an issue. The underlying network infrastructure should have mechanisms to utilize and enhance the protocols that run on it. The naming scheme for data objects is to be considered one such protocol. Thus the architecture should be able to support and enhance it.

This dependency is paramount when one observes the literature in the naming schemes. In most comparisons, one analyzes the architecture along with the naming schemes and always gives a joined result. In this work, we avoid doing the same. We look at the benefits that a naming scheme can give on its own and then consider its benefits and drawbacks since the underlying architecture can be fine-tuned based on the naming scheme chosen. However, this section is to highlight the fact that there is an interconnection with the architecture as the naming scheme itself cannot do much alone. It needs the support of multiple mechanisms for functioning accurately.

2.2.6 Scaling of Names

Scaling of names is a very big challenge that any naming scheme faces. The number of data objects on the Internet is very large and being able to represent such a large amount of data uniquely is a very daunting task. The naming schemes may be able to perform well when the number of objects is small but being able to scale to be able to cover the whole of the Internet and more is a very challenging task. It is desirable that the naming scheme in question be able to keep up with the huge amount of data already on the Internet and be able to adjust to the enormous amounts being added almost daily and keep doing so for many years to come. If one is shifting to the ICN paradigm of networking, this is a very important challenge since if not fulfilled, one cannot successfully migrate to the ICN paradigm without experiencing a degradation in the network performance [10].

2.3 Requirements for a Naming Scheme and Performance Metrics

There are various requirements for an efficient naming scheme for ICN. These parameters are divided into two categories: primary parameters and secondary parameters. Primary parameters are must-have parameters for any naming scheme while the secondary parameters are desirable features for a naming scheme and can be used to evaluate and compare different naming schemes.

2.3.1 Primary Parameters

Identification: The naming scheme must be able to provide a unique name to all unique content. It is desirable that the network be able to establish one-to-one mapping of the data available in the network to that which is being requested by the user. If there is a unique name that identifies each data object, then there is no confusion in the network and the correct requested data will be returned to the user [34].

Assistance in routing: The naming scheme needs to assist in routing information by helping make the content easily reachable. In the ICN paradigm, since data objects are directly queried rather than querying the servers that are hosting them, the data objects need to be routable and identifiable by the routers themselves rather than only the server and the client [29,31,34].

Human readability: The naming scheme needs to be human readable. The user is searching for content based on its name and should be able to identify and remember the name easily without having to remember some random characters as a name. This is one of the most debated requirements in any naming scheme proposal. It is required for the user to identify the data object that he or she queries for. It is interesting to note that it does not mean in any manner that the user should be able to read the whole name but that the user should be able to read a portion of it that conveys the required information.

Security: The naming scheme should be able to certify the content (either through self-certification or a strong binding that is attestable by a third party). It is essential to satisfy the security requirement of a naming scheme. It should be possible for any user to authenticate the data object that has been returned by the network. This certification ensures that data objects are not forged or corrupted in any manner [34].

Scalability: The naming scheme should be scalable. This is one of the most important aspects of any naming scheme. As mentioned previously, in today's world, there is a lot of data available and it is continuously growing. Any naming scheme we use should be able to scale to cover all this data and more. The naming scheme in question should have the namespace required to name all the data objects and help locate them in a large network. If it cannot scale then the naming scheme itself does not seem to have much value beyond a theoretical exercise since no matter what the properties, it becomes impossible to use it in a real-world scenario.

2.3.2 Secondary Parameters

These are the secondary parameters used to evaluate the various naming schemes:

The ease of name resolution: This is the mechanism by which a name can be resolved to some content, i.e., an object can be identified by its name. At first glance it seems to have a similar type of functionality as a DNS. It seems intuitive that there has to be some sort of storage/mapping involved for such a function that binds the data object with a name for the object and that is generated in a specific manner depending on the type of naming scheme used. This parameter determines how difficult it will be to perform name resolution on each of the protocols mentioned [29].

Implementation complexity: This may decide the fate of any proposed naming scheme. Any naming scheme cannot exist in isolation and must coexist with an existing TCP/IP network for a long time to come. Thus, it should not be overly complex to implement and should be able to work along with the legacy network for a given time frame until a complete shift is done to the ICN paradigm [30].

Effect on caching: This is a very important and interesting criterion. It is assumed that the concept of caching is being utilized to improve the efficiency of ICN. Research work is being carried out today to move in that direction by deploying various caching schemes and improving their efficiency or usability in the context of ICN. The reason for its relevance here is that any object that is cached will need to be identified and this is done via naming. It is desirable that there is a property that is being provided by a naming scheme that can be used to better handle caching, or at the minimum that the naming schemes do not impact caching techniques adversely even if they do not provide any benefit to the caching schemes being deployed. Thus, an implemented naming scheme should aid in caching functionality and other smart data processing by the network.

Effect on existing names: IP address can be visualized as the naming scheme in the traditional network. Since the ICN naming works on data whereas the IP works on devices, there is no apparent reason for there to be conflict between the two naming schemes. However, if there is another naming scheme implemented (e.g., a hierarchical naming scheme in the form of URLs), then this naming scheme should not cause problems when interacting with existing naming schemes. It should be able to coexist with other naming schemes.

Ease of translation from one naming scheme to another: This is another important factor. Considering that ICN is in the initial stages, there is a high probability that multiple naming schemes might exist in the network at the same time. This parameter talks about ease of translation from one naming scheme to another one to promote easier interworking between different naming schemes. Thus, a naming scheme that can be translated easily to other naming schemes is preferred over one that cannot be translated easily. This becomes an important issue because different naming schemes have different properties that might make them the preferred naming scheme for a specific real-world situation. Thus, a different naming scheme may need to be implemented on a different network and in such a situation, the lack of interoperability should not be a hindrance.

Ease of addition of already existing content: This is a very important factor. In today's network, there is a huge amount of content that is already uploaded. It cannot be taken off. Thus, the naming scheme should be able to work with these objects and add them without too much complexity. This is also one of the challenges in introducing the ICN architecture in today's network.

At the end of this discussion we would like to add that to the best of our knowledge, the naming schemes being proposed are in a very theoretical space and the practical work regarding simulation or modeling has not progressed very far in the field of naming. Various simulators have adopted a naming scheme and worked with it but there is very little work being done on comparing the performance of various naming schemes through a performance evaluation. Thus, most of the content discussed so far and to be discussed is based on various literature on naming schemes as well as other metrics and ideas discussed theoretically [5,30].

2.4 Types of Naming Schemes

2.4.1 Attribute-Based Naming Scheme

2.4.1.1 Description Attribute-based naming schemes are based on the assumption that any data object that is uploaded will have some key attributes that can identify it. These attributes might be user provided or may be an attribute of the data object itself. The naming scheme makes use of this property by making a list of attributes and then choosing a name that is a combination of the attributes that identify the data object. Thus, the names derived from this naming scheme usually are a set (ordered or not) of what are considered defining attributes in a specific instance. These attributes are then merged together in a defined protocol to form a name. The actual protocol to form the name can vary from something as trivial as writing the attributes one after another to form one long string to more complicated protocols that use separators and have spaces for specific attributes. A form of naming based on the principles of attribute-based naming is mentioned and used in the CBCB architecture [9].

2.4.1.2 Performance on Primary Parameters

Unique identification of content: Fulfilling this property is a challenge for the attribute-based naming scheme. There is a huge variety of data and a large amount of it. Analysing all that data to gather a list of attributes that make each data object in that collection of data unique is a very daunting task in and of itself.

Furthermore, it is not the case that there always exists such a set of attributes that can guarantee the uniqueness of each data object. This makes the challenge even more difficult as when we begin this endeavour, we cannot guarantee that we can identify content uniquely based on a set of attributes. This becomes a very big challenge to the attribute-based naming scheme, as identifying content uniquely is a very desirable trait, one that attribute-based naming is unable to guarantee at all times.

Assistance in routing: Attribute-based naming schemes generally cannot assist much in routing. The names in attribute-based naming are formed using various attributes of a data object. This may or may not include the originating server. The attributes themselves are part of the data object and do not have any knowledge of the network and cannot convey the same to the network. This kind of naming does not in any way interact with the data object's location in the network or utilize other knowledge of networks. It also does not possess any of the inherent properties that other naming schemes possess like utilizing the tree like structure in hierarchical naming. This naming scheme will need a name resolver with a different name that is routable to be able to assist with routing, which just adds to the complexity.

Human readability: This is the part where the attribute-based naming scheme shines. It has been long observed that most often when people search for data objects, they search using some specific keywords. They query the object on a search site using certain keywords which are otherwise descriptors of the objects that they know of. The attribute-based naming takes advantage of this feature. The attribute aspect of the naming is very similar to the keywords and it may be the case that some of the keywords are attributes themselves. This makes the Named Data Object (henceforth NDO) very searchable in terms of human understanding and makes the name human readable since it is something that a human can relate to.

Security/content self-certification: This requirement is not necessarily satisfied by the attribute-based naming mechanism. The idea on which the attribute-based naming mechanism is based has evolved to utilize the keyword search pattern observed in users' search of sites and help identify NDOs with a naming scheme built along the same lines. This does not in any way guarantee security or that the object in question is the correct object. That would require other mechanisms that are extra and outside the scope of the philosophy of the attribute-based naming mechanism. While the naming scheme does not in any way seem to work against security, the core philosophy of attribute-based naming also does not seem to have provisions for security or validation of the integrity of the content.

Scalability: This is the aspect where the attribute-based naming faces huge challenges. The naming scheme in and of itself is not scalable. Considering the amount of data on the Internet today, even generating a list of attributes and a rule that guarantees uniqueness for each individual data object is an extremely difficult task. Furthermore, it is not the case that the list of attributes found will be a small list and the list may contain a huge list of attributes, which can lead to very long names. Long names are undesirable as they impact time taken to cache and require more processing and memory in comparison to shorter names thus becoming a possible bottleneck in the network. While it is possible to utilize attribute-based naming for a network containing a small number of objects, creating the list of attributes and then uniquely naming the data objects becomes an extremely difficult task as the number of data objects increases.

2.4.1.3 Performance on Secondary Parameters

Name resolution: The attribute-based naming faces challenges in this particular requirement since it cannot guarantee uniqueness in identification of data objects. One can identify an object with the attribute-based naming scheme but as stated before, to ensure that the object is uniquely identifiable is a very difficult task. This makes name resolution a difficult task since the name resolver cannot be certain that the name points to the correct data object as it may be the case that there are two or more data objects with the same name. This facet also makes it hard for the user since the user now may be provided with a NDO which he or she has not requested.

Implementation complexity: This aspect has been discussed to a certain extent previously and can be observed in the various sections of the primary requirements. Previously we have stated that it is hard to ensure uniqueness from a list of attributes. Let us explore that in a little more detail. First, it is no secret that there is a lot of data on the Internet. And the data is of different types having different attributes. The first major problem would be to ensure that all the types of data on the Internet have their attributes listed and the attributes to be considered in the naming scheme identified. Those that are familiar with concepts from data analytics will appreciate how difficult a task this is considering that the size of the dataset is immense. The analysis of this data to generate a list of unique attributes is estimated to consume a huge amount of time and processing power. Now after this analysis is done, we need to come up with a set of rules for uniquely generating a name from among the set of attributes identified as key attributes for each kind of data object. This is also a hard task considering that while outliers are present and allowed in data analytics, they are not permitted here. Apart from this, other aspects also need to be considered like the difficulty in building a name resolver that can handle problems highlighted in the name resolution section, the difficulty in building a routing mechanism for networks using this naming scheme considering that the names themselves do not aid in routing in any significant manner, etc. All the previously mentioned aspects make implementation complexity very high which while undesirable cannot be easily avoided in this case.

Effect on caching: The attribute-based naming scheme does not seem to have provisions to aid in better caching. There is no inherent property that can be used to affect caching policies in a positive manner. One property of naming schemes that may affect caching is the name length, which is the size of the name in desired units. Constant name length would be desirable over variable name length as variable name length poses a greater challenge to the network with respect to effective handling. In the attribute-based naming scheme, it is hard to guarantee that all the different types of data have similar name lengths. Assuming that the name is formed from a certain set of attributes, it does not tell us that every data object will possess all the attributes in the set in a manner that is not NULL. This makes ascertaining the name length a difficult task thus causing variation in name length. This may adversely impact efficient caching.

Effect on existing names: The attribute-based naming scheme is not very compatible with the other existing naming schemes. The attribute-based naming scheme does not have any knowledge of the network or a certifying authority. It utilizes only the attributes of the data object. As previously mentioned, other mechanisms are required for the additional details like a routable name. Since it needs the support of other mechanisms to locate NDOs and runs in conjugation with those mechanisms, we can assume that those mechanisms can be implemented to exist with the existing names if needed.

Ease of translation: The attribute-based naming scheme in and of itself does not work with the other naming schemes mentioned here since it does not have a certification in the form of a hash string (which can be an attribute but seems redundant) which can be used with w.r.t. flat naming; neither does it have the hierarchical structures that are looked for in hierarchical and hybrid naming schemes. This makes translating between the different naming schemes and attribute-based naming a hard task.

Ease of adding pre-existing content: This is discussed to a certain extent in a previous point. As mentioned, there is a lot of data today. Identifying the amount of types of data, listing their attributes, performing analysis, and identifying a set of key attributes that can guarantee a unique name is hard. Once done, generating that unique name for each of the NDOs is easy in comparison if the previous part holds true, since we would have derived a rule. It is always going to be hard to add every data object on the web today into the new ICN network. The attribute-based naming scheme does not contribute in any manner to that difficulty once the rule is generated. All one has to do is generate a name for the data object according to the name generating rule and add it. The only complication that is caused due to attribute-based naming is to generate the rule and this is a difficult task as previously discussed.

2.4.1.4 Summary The advantage of the attribute-based naming scheme is that it utilizes the observation of querying for objects using keywords and attempts to generate a name using the same principle. This attempts to make queries for the data object more efficient and makes the object more human readable.

When one compares this to the requirements stated above, we observe that the naming scheme handles the readability requirement well but does not fare well in the security, uniqueness, and the scalability requirements. It also does not assist in the routing by giving useful information and requires a third party to help it resolve names and route.

It has a lot of advantages but it does not satisfy many of the basic requirements of a naming scheme and requires additional complexity along with major changes to satisfy them. While that makes implementing this naming scheme a somewhat difficult challenge, it is important to note the fact that this naming scheme does have some properties that are desirable [9,12,23,24,27].

2.4.2 Flat Naming

2.4.2.1 Description The flat naming scheme is a group of naming schemes that currently are the subject of a lot of focus in the space of ICN naming. A lot of research is being conducted where the focus is on the advantages the flat naming schemes can bring as well as how to make them scale and assist in routing. A lot of flat naming schemes vary in actual policy regarding which hash function to use, how the one-to-one mapping is achieved, etc., but flat naming schemes generally work in the following manner. A string of characters is generated as the output of a hashing function which is bound to an NDO. This string/NDO combination acts as a key/value pair for a hash table.

This string is then considered the name of the NDO. Depending on the generation process of the name, self-certification and validation can be enforced. We highly encourage you to read up on the current work in this field because it is an emerging area and the work is happening live [9].

2.4.2.2 Performance on Primary Parameters

Identify content uniquely: The flat naming scheme's name generation mechanism, which involves using a one-to-one mapping between a string generated by a hash function and an NDO, ensures that each NDO has a unique name that is generated by the generating authority. One of the challenges to consider while generating this name is to fix the name length as this impacts namespace. Please note that this does not involve security concerns where someone else uses the same name for a different object. All the flat naming scheme assures us is that it is possible to identify content uniquely using their names due to the existence of one-to-one mapping between the NDOs and their names.

Assist in routing: The ability to assist in routing is a challenge to flat naming schemes. The naming scheme by design is flat which implies that there are no hierarchical properties that one can use to assist in routing in any manner. The name generation mechanism also is using a hash function, which need not contain any knowledge regarding the location of the object in the network. Enabling some form of routing assistance in flat naming schemes or creating routing mechanisms which can work on ICN architectures which utilize flat naming schemes is one of the focuses of naming research today. The flat naming scheme by itself does not offer any assistance to routing.

Human readable: The flat naming scheme again by design is a string of characters generated to have a one-to-one mapping between the string and an NDO. This design does not and is possibly incapable of considering human readability because the string generated may also contain special characters which are not in the alphabet or numbers.

Generating a string that can help security and uniqueness affects the readability aspect of the naming in an adverse manner. This causes names generated in the flat naming scheme to be unreadable and possibly misleading as well for humans.

Certify/validate the content: The flat naming scheme allows for certification or validation of content. This is done by the choice of the hash function which generates the names. Naming schemes can be validated by trusted third-party authorities that generate unique names for data objects. Choice of the hash function in such a naming scheme and the presence of the trusted third party allows the names to be self-certifying since the name can be checked at any time to ensure that the object received is the one that the name points to. It is to be noted that the mechanism to implement this depends on the kind of mechanism used for name generation but it does not seem to be a very big problem and there is a lot of literature available where the validation aspect of the naming schemes has been considered.

Scalability: Flat naming schemes face a challenge in terms of scalability. While an appropriately chosen name length and the huge number of characters available ensure that namespace is not an issue, there is no provision for locating an object and routing. With a flat naming scheme, it becomes difficult to assist in locating the correct data object. Implementing flat naming schemes for a small network is perfectly possibly since routing based on flat names is possible for a small number of data objects but as we start to scale, getting desirable results becomes extremely challenging.

2.4.2.3 Performance on Secondary Parameters

Name resolution: The name resolution mechanism in the flat naming scheme does not seem to be complicated since the naming scheme ensures that each object is uniquely identifiable and that the name length is not very large. So long as there exists a third party that can perform the translation, name resolution does not become an issue and only involves contacting this third party and asking for the location/characteristics of the data object that a name points to.

Implementation complexity: The flat naming scheme has components that are challenging to implement and other aspects that are not so challenging. The creation and existence of a trusted third party who helps in ensuring certification and validation of content is a daunting task. This impacts the naming scheme because it is this third party that allows for and monitors the hash function that is used to ensure uniqueness in names generated. Assuming that this third party exists, assigning names to the data objects today becomes a somewhat simpler although time consuming task. One of the greater challenges when implementing a flat naming scheme is the lack of assistance in routing, which is still unsolved.

Effect on caching: The flat naming scheme does not have provisions for caching but also does not seem to adversely impact caching mechanisms. Two of the problems that naming schemes can contribute to w.r.t. caching policies are related to the length of the name and the variation in lengths of names. A large name length is not desirable but this depends on studies that are yet to be performed regarding optimum name length size and whether that name length specified has sufficiently large namespace. The variation in the name length on the other hand is a challenge that is more efficiently handled by the flat naming scheme. This enables it to not adversely affect caching in ICN.

Effect on existing names: The flat naming schemes affect the already existing names neither favorably nor adversely. These schemes can replace the names easily since those names need not be considered to give a flat name. Also, the flat name generated can possibly be part of the other naming schemes thus making it easier to coexist but if this is not the case, then interworking mechanisms for the naming schemes need to be considered since the flat name cannot effectively communicate with the other naming schemes.

Ease of translation: The translation aspect of a flat naming scheme seems to be somewhat straightforward with the presence of a third party for name generation and resolution. The flat portion of the name generated can be added to the other names themselves (this is one of the ideas behind the hybrid naming schemes), thus enabling translation between naming schemes, but the flat naming scheme does not have any hierarchical properties or any indicator of the attributes of the NDO in the name itself that can be understood by the other naming schemes. This implies that it is easier for other naming schemes to be translated to a flat scheme as opposed to a flat scheme being translated to another naming scheme.

Ease of adding pre-existing content: This is the easiest aspect of the flat naming scheme. Technologically, it is not very difficult to add pre-existing content. Once work has been done to decide on a name length and the third-party authority is set, all that needs to be done is to route the data object to the third party so that it can be given a name that is generated. This process is very time and resource consuming but has to be done and the same is true for any naming scheme considering that to add the pre-existing data, the data has to first be given a name which is a time-consuming task. The flat naming scheme in and of itself does not add any considerable difficulty to adding pre-existing content [6].

2.4.2.4 Summary The flat naming scheme satisfies some of the parameters while facing challenges in certain other parameters. It is the naming scheme used in most ICN architectures today in the form of DONA, PSIRP, and NetInf and has some very interesting properties that are useful. Work is being done to lower the negatives in an attempt to make this the naming scheme that is most favored in ICN due to the huge number of benefits it offers in terms of uniqueness and security [9,13,14,16,19,23,27,32].

2.4.3 Hierarchical

2.4.3.1 Description Hierarchical naming is a naming scheme that takes advantage of the tree-like structures that hierarchical structures possess. There is a hierarchy or a parent-child kind of relationship between the various nodes which can be taken advantage of for various purposes. One of the examples is the use of IP addresses. IP addresses are hierarchical in nature so that they can be clubbed together and only the prefix used to identify a network.

The hierarchical naming scheme being discussed resembles the structure of the URLs that are currently used today. The terms used in the name can be based on a variety of properties like server location/geographical location. While this gives huge advantages in terms of routing by partially emulating the prefix aggregation property of the IP addresses, it fails in a few aspects. The structure of the name is as follows. Each part of the name is separated by a separator ("/" in the case of a URL). There is no limit to the number of levels it needs to go in terms of depth.

A hierarchical naming scheme is currently used in the NDN simulator. The names and the methods used to generate these names are application- or application developer–dependant with some guidelines being given by the makers of NDN. For those interested, we suggest a closer look at the NDN architecture and the naming scheme used there. The scope of this chapter covers discussing the properties displayed by hierarchical naming schemes that can possibly be used for ICN.

The benefits of this naming scheme and its strengths are that it can take a very distributed approach. It does not require a third party and once the guidelines for creating a name are fixed, then each system can generate data objects and their own names for those data objects. Thus, it becomes important that the guidelines chosen to set names be observed and studied extensively to gain maximum benefit [9].

2.4.3.2 *Performance on Primary Parameters*

Identify content uniquely: In the hierarchical naming scheme, it may be possible to identify data objects by generating unique names for them depending on the guidelines that govern name generation. An example would be to add the time of generation along with location of generation of a data object to its name. Another idea would be to add the name of the publisher and owner of the object in its name along with an identifier that the publisher generates to uniquely identify his objects. There may be many such guidelines that can help generate a unique name, while there may also be guidelines that do not guarantee a unique name. These approaches need to be studied in greater detail because while it may be possible to generate unique names, this has to be proved before a particular guideline is utilized.

Assist in routing: The hierarchical naming scheme is a naming scheme that can provide assistance in routing. The hierarchical property of the name helps in aggregation and efficient pruning of the routing table for more efficient usage of memory in the router as well as more efficient routing. This can be used for identifying the names, for clubbing names based on prefixes, etc. The hierarchical naming scheme can assist in routing to a greater extent compared to naming schemes due to the tree-like nature it possesses. This is again very similar to the IP-like addressing scheme since the IP addressing itself is somewhat hierarchical in nature. Again it is technically possible to come up with a hierarchical name that does not assist in routing but considering that assistance in routing is a desirable property, such a hierarchical naming scheme would be of little value to us and thus be discarded for better schemes [31].

Human readable: This naming scheme may be readable depending on whether special characters are used as part of the names, the length of the names, etc. It is very similar to the URLs being used now and is easily readable by humans in most cases. Although an implementation where it is not human readable is technically possible, it is generally not the case. Again, this depends on the guidelines. If human readability is deemed to be an important property, then the guidelines will reflect this. There is nothing in the naming scheme that seems to prevent human readability.

Certify/validate the content: Hierarchical naming schemes do not provide mechanisms for certification or validation of content. This becomes apparent in the naming scheme itself considering that the naming scheme is distributed and does not have any certifying authority. One of the greatest strengths of the hierarchical naming scheme, which in this case is the distributed nature of its naming and its developer-dependant naming, becomes the reason for it to lack validation of content as it is difficult to validate content unless a form of validation is agreed upon. In the ICN paradigm, it falls upon the data to self-certify itself and the hierarchical naming scheme is unable to do so.

Scalability: The hierarchical naming scheme can possibly ensure that content names are not repeated and that the namespace is sufficiently large based on the rules used to generate the name. At the same time, the hierarchical naming scheme also faces many challenges for scalability. While from a purely naming perspective, it may seem that the properties of prefix aggregation, large namespace, etc. are properties that can be useful for scaling, they are also time consuming due to overhead and other problems when faced with a large number of data objects. It may well be the case that the hierarchical naming scheme fails to work efficiently in real-time constraints. Thus, as far as scalability is concerned, from the naming perspective the hierarchical naming is scalable but from the whole of ICN where naming plays an important role, the hierarchical naming scheme has a lot of challenges to overcome [3,4,18].

2.4.3.3 Performance on Secondary Parameters

Name resolution: In a hierarchical naming scheme, since the names are the responsibility of the content provider with some generic guidelines given by the network, name resolution can be both easy and difficult. Name resolution can be simplified based on the guidelines set to create the name depending on how the creation of the name utilizes the tree-like structure of the naming scheme. For example, consider a hierarchical naming scheme that gives names along the lines of the location (as given in Figure 2.3).

If the naming scheme guideline states that the name of the content should be based on the upload location—like if one is posting from Bangalore, electronic city, Infosys, the name should be "Asia/india/bangalore/electronic-city/company/serverx/content-name/part-x," with an upper limit on the size of each fragment—then the resolution becomes very easy

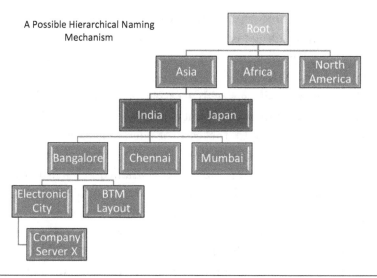

Figure 2.3 A possible hierarchical naming mechanism.

since in the search, one need not search for content that is not in this particular subtree of the naming scheme. But at the same time, this particular scheme also causes problems in resolution because of the fact that the name size is variable. Thus the size of the data tables that store a name cannot be decided beforehand. Also in such a hierarchical mechanism, it may be the case that the property on which the hierarchy is built may not be easily resolvable for all kinds of data. For example, in this case, a user requesting the data from the United States does not know where the data is. The data may have been uploaded from the given location, but it may be hosted elsewhere. This causes problems at the level of the resolver, at the level of storage for caching, etc., since long variable length names are not desirable. Also, as stated previously, uniqueness is an issue here, which might cause further issues with name resolution [18].

Implementation complexity: The hierarchical naming scheme faces various challenges that increase the complexity of implementation. One of the challenges is name length. The technology today cannot handle a name with variable name length with minimal delay. The second challenge is that there is a lot of overhead involved in parsing each portion of the name to ensure effective routing. This is also a very time-consuming process. The third challenge is to come up with a set of guidelines for naming that can ensure unique names, i.e., a one-to-one mapping between data objects and their names. There may possibly be more challenges with w.r.t. overhead etc., but we believe these to be the most important challenges currently [18].

Effect on caching: In the case of hierarchical naming, the naming scheme might cause a lower number of objects to be cached due to large sizes. As mentioned previously, work needs to be done to check whether the naming scheme implemented affects the caching at a much deeper level than has been observed so far in terms of performance, but a larger name generally implies larger tables for object names which implies larger consumption of memory for the same number of objects as well as greater overhead. Large names also impact the amount of data transmitted in packets because while the MTUs are constant, if more space is taken up for large names, then less data can be stored and transmitted in each packet, thus increasing bandwidth consumption.

Effect on existing names: The hierarchical naming scheme does not contain any mechanisms that affect coexistence either favorably or adversely. To be able to coexist with other naming schemes like flat naming schemes or attribute-based naming schemes, it is the case that interworking mechanisms, while challenging to provide seem to be easier done in the case of hierarchical naming schemes due to the lack of strict guidelines in the name generation.

Ease of translation: The translation aspect of a hierarchical naming scheme is a somewhat challenging task. The hierarchical naming scheme need not possess a set of attributes that can uniquely identify an object thus making it difficult to translate it into an attribute-based naming scheme, nor does it need to have a self-certifying string, making it difficult to translate into a flat naming scheme. It also cannot be figured out from the other two naming schemes since they do not possess any of the properties that a hierarchical naming scheme requires thus requiring the intervention of a third party to help translation in any manner. Although it may be possible to coexist, translating easily from one naming scheme to another is a daunting task.

Ease of adding pre-existing content: In the hierarchical naming scheme, there lies a question of which guideline covers the content. The content today is uploaded in multiple servers and each user might follow his or her own style of naming which might make it difficult to differentiate between the content. There might be some challenges related to the addition of already existing content in hierarchical naming but it is more a problem of naming general content in the hierarchical naming scheme than of naming older content [8].

2.4.3.4 Summary The hierarchical naming scheme performs well on many of the basic requirements while it does not fare well on others. This naming scheme has its strength in the fact that it can assist routing. This is the naming scheme that is implemented on one of the most popular ICN simulators, NDN. There seem to be a lot of abilities that are offered by the hierarchical naming scheme which are similar to the current identification, which is IP addressing. These make it a very strong contender for being used as the naming scheme for ICN. Yet it lacks various abilities like self-certification and security. Nonetheless, the hierarchical naming scheme is used a lot in various ICN contexts and although many new simulators use a flat naming scheme, NDN continues using the hierarchical naming scheme and works with it [3,4,11,15,17,23,24,27,28].

2.4.4 Hybrid Naming Scheme

2.4.4.1 Description This is a naming scheme that works with a school of thought that questions whether it will be beneficial to take the best properties of the other naming schemes and club them together. If it is then what would be the possible naming schemes that are to be clubbed together? It seems to be more logical to try to merge them in some fashion to leverage the advantages of both. Have a flat part of a name for self-certification as well as a hierarchical part to make use of the aggregation property to a certain extent. One can also use the attributes as part of the name to make the name more user-readable.

The hybrid naming scheme is a class of naming schemes that attempts to leverage the three previously proposed naming schemes. It may contain two or three parts. This work assumes that it contains all three parts and based on the various naming schemes proposed later, the naming scheme can be modified.

The first part of the name is the hierarchical part. This part follows the hierarchical naming scheme in that there is a hierarchical structure that is being exploited in the naming. But this is not necessarily left to the discretion of the application or the application developer. There might be a central body that is governing this allotment. The second part of the name is based on the flat naming convention. This uses encryption and public key concepts to generate a random string which is part of the name. The third proposed part uses concepts of attribute naming. In this part of the hybrid naming scheme, some key attributes that are useful to identify that object are added as part of the name of the object. The naming scheme uses delimiters like the hierarchical naming scheme to separate the different parts of the name as well as the different parts of each naming scheme itself [2].

2.4.4.2 Performance on Primary Parameters

Identify content uniquely: The hybrid naming scheme fulfils this particular requirement. The idea of having a flat portion of the name help with two things: (1) unique identification of objects and (2) self-certification. This naming scheme, although having a longer name than that of a flat naming scheme, is still useful since we require only a portion of the name to uniquely identify the object. This gives it an advantage that is like that of the various flat naming schemes proposed for the purposes of unique identification.

Assist in routing: The hybrid naming scheme seems to fulfil this requirement to a certain extent. The presence of a hierarchical portion in the hybrid naming scheme can be theoretically used for routing assistance. Hierarchical names as we already know help in routing to a certain extent due to the aggregation property which can lead to easier identification and smaller routing table requirements. The hybrid name, although not containing the full hierarchical name, does have a hierarchical portion which can also use the aggregation property to a certain extent and that helps it in routing. It may not be as effective as hierarchical routing and routing techniques developed need to be a combination of both hierarchical and flat routing techniques.

While it is the case that hybrid naming schemes can assist in routing to a certain extent, it is also a fact that the hybrid naming schemes tend to have long names. Although the name length may not be variable to the level of hierarchical naming, it is still somewhat variable as well as long. This is not desirable as long names of variable length affect routing adversely. Some measures have been proposed on how to possibly avoid this but they are beyond the scope of this chapter.

Human readable: The hybrid naming scheme faces some challenges w.r.t. the human readability aspect of the name. Considering that the name consists of three parts and the flat naming portion does not contribute to readability in any manner, the readability of a hybrid name is based on the hierarchical naming– and the attribute naming–based portions. The hierarchical naming portion covers only the first few levels of the hierarchy to enable aggregation. This implies that it does not contribute much to the actual data object in question as opposed to the group that the object belongs to in some manner. Thus, the actual human readability aspect is dependent on the attribute-based naming. This approach is valid but due to the possible long length of the various hybrid names, this quality of human readability may decrease depending on the positioning and the visibility of the attribute-based name portion as well as the attributes chosen for the attribute-based name portion.

Certify/validate the content: The hybrid naming scheme utilizes the flat naming portion of its name to help certify and validate content. The benefits are much the same as a flat name and it works in a similar manner. Private key encryption and public key decryption help certify the publisher as genuine and the hash with the self-validation name helps to certify that the data object has not been tampered with, although this requires the intervention of a trusted third party who maintains records of the keys used. This is similar to the problem with the flat naming scheme.

Scalability: In the hybrid naming scheme, there are three components. The first is the hierarchical part, which is used to select which portion of the names to search in. In this part, the length of the hierarchical part should be decided such that it does not adversely impact the naming scheme. It should not be the case that due to a vast number of objects, there is no significant benefit from the aggregation ability of the hierarchical naming scheme and the flat name search is conducted on a huge number of objects.

The second part of the name is the flat naming scheme. Again, the string of characters should be sufficiently long to offer enough namespace for the various data objects provided. It seems to be scalable but to what extent is a question that cannot be answered at this point in time. But there definitely is an upper limit to the number of names that can be uniquely generated for a set string length. This needs to be considered when choosing an appropriate length for the flat name part of it.

The third and final part of the hybrid naming scheme is the attribute-based naming scheme. This is the part that is the least challenging since the attributes of an object are used for human readability and to possibly aid keyword searches and thus do not contribute to scaling challenges in a significant manner.

2.4.4.3 Performance on Secondary Parameters

Name resolution: Hybrid naming has three different segments. Thus the code for the name resolution will be comparatively complex when compared to other naming schemes. First there should be a portion of the resolver that can search using keywords and present them to the user. The uniqueness aspect of the name resolution as well as variable length is not an issue since the flat name part of it can be used uniquely for name-resolution. The name resolver can employ a policy used for hierarchical naming to prune a segment of the name tree. Obviously, we cannot complete it based only on the hierarchical naming scheme since it is only a part of the name and not a complete hierarchical name. Once the resolver arrives at a particular branch of the list of names, then it uses the mapping that is used for the flat naming scheme portion of a name. To the best of our knowledge, practical simulations have not been performed regarding which is better—i.e., the hierarchical name resolution mechanism, the flat name resolution mechanism, or the hybrid name resolution mechanism—but from a functionality point of view as well as a scalability and implementation point of view, the name resolver for hybrid naming, while taking more effort in its construction and having overhead that seems to be comparable to the hierarchical naming scheme, seems to be the better choice since the advantages outweigh the disadvantages.

Implementation complexity: The various challenges faced when implementing a hybrid naming scheme have significant contributions from two major aspects. The first is challenges faced in the implementation of the various naming schemes that make up the hybrid naming scheme, i.e., challenges due to the implementation of the hierarchical naming scheme, the flat naming scheme, and the attribute-based naming scheme. The second major aspect comes from merging the various naming schemes and getting them to work seamlessly.

This is not a clean slate implementation. This is a new paradigm that should coexist with the traditional paradigm, which is a daunting task for this naming scheme. It requires a significant amount of infrastructure changes as compared to other naming schemes because it requires a centralized body that governs decisions, unlike the hierarchical naming. It also requires a trusted third party to keep track of the keys to ensure self-certification.

This naming scheme also requires a lot of processing overhead in terms of switching between naming schemes, which is not present and thus is not a factor in a pure naming scheme. Thus, although it can be implemented and gives us a lot of advantages, the cost is a complicated network infrastructure which needs to be planned carefully.

Effect on caching: In the case of hybrid naming, the variation in the name length may be low depending on the naming scheme created, which is desirable. Also, considering that searching among all the data objects present is not required, but only among the data objects that are cached, which is a significantly smaller subset of all the data objects, it should be feasible to store only the flat name part of it. This helps to further reduce the length of the name while still retaining the uniqueness aspect of the name. This might imply that a greater number of names can be cached without severely impacting the performance.

Effect on existing names: The hybrid naming scheme does not have any mechanisms that impact any existing naming schemes either favorably or adversely. The hybrid naming scheme is complex to implement. If there are already existing names, then it becomes challenging for the hybrid naming scheme to effectively work with them. Already existing names can be converted into hybrid names with some effort. This conversion is a more complicated function when compared to the other naming schemes due to the generation of two different missing names (assuming that the attribute-based naming consists of all three naming schemes). This leads to overhead. This along with implementation complexity makes it difficult for a hybrid naming scheme to exist with other naming schemes.

Ease of translation: The hybrid naming scheme allows for translation between the various naming schemes in a manner that is less challenging when compared to the other naming schemes. This can be attributed to the fact that it possesses all three kinds of naming schemes within its name. Let us consider the translation from a flat name to a hybrid name. The first hierarchical part can be filled with a special character and the flat part can be the flat name proposed. The inclusion of a special character in the hierarchical part directs the name resolver to check the store that handles that particular naming type. Figure 2.4 shows this exact scenario.

In this figure, ";" is used to separate the different attributes while "\" is used to separate the hierarchical part of the name. The symbol "~" is used to differentiate between the different

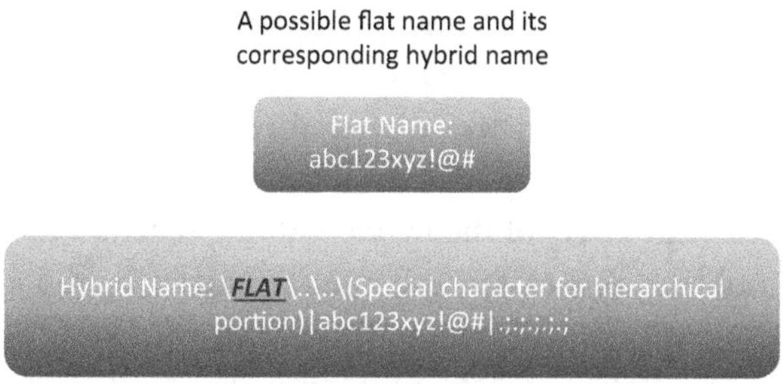

A possible flat name and its
corresponding hybrid name

Flat Name:
abc123xyz!@#

Hybrid Name: **_FLAT_**\..\..\(Special character for hierarchical
portion)|abc123xyz!@#|.;.;.;.;.;

Figure 2.4 A flat name and its possible correspondent hybrid name. (From Zhang, H. et al., "Uniform information with a hybrid naming (HN) scheme," *IETF Internet draft draft-zhang-icnrg-hn-07, Work in Progress*, October 2017.)

parts of a possible hybrid name. While this forces us to relinquish the prefix aggregation property of the hierarchical naming scheme, it saves us the overhead of generating a hierarchical and attribute-based name.

Translating from an attribute-based naming scheme to a hybrid naming scheme seems to be more challenging since it is possible that the NDO cannot be uniquely identified. This is possible due to the fact that hybrid names may have an attribute-based component in their naming that can be used to help ease the translation between attribute-based naming and hybrid naming. Again, it is reiterated that this procedure without generating the hierarchical and flat naming components of the hybrid naming does not give the benefits that are gained from utilizing hybrid naming.

A translation from a hybrid naming scheme to a hierarchical one is also possible where the flat id is treated as an identifier attribute and is only added in the hierarchical name as part of the name itself. This has already been proposed in [2]. Converting from a hierarchical name to a hybrid name is a more significant challenge, however. Although the hierarchical part is taken care of by the hierarchical naming scheme, it is to be noted that the hierarchical portion of the hybrid name is fixed to a certain depth which may not capture the complete hierarchical name. This along with the lack of any flat naming components as well as attribute-based naming components causes a problem in identification of the object. The generation of a flat component to identify the object is cause for additional overhead. The attributes, while important for the sake of user readability, can be forgone to a certain extent but the flat naming component is required due to the self-certification and uniqueness property of the hybrid naming scheme.

Ease of adding pre-existing content: In the hybrid naming scheme, each piece of content has to have a unique id attached to it. Also, attributes have to be provided by the provider of the content and tacked on for the hybrid name. Apart from this, the hierarchical portion of the name also needs to be handled effectively. These names require extensive cooperation between multiple bodies as well as slow replacement/ modification of the already existing content to include the name given to it. Because of the above reasons, it will be more resource consuming and challenging to add pre-existing content to an ICN architecture with the hybrid naming scheme when compared to other naming schemes [2].

2.4.4.4 Summary Hybrid naming in theory seems to be able to leverage the advantages of both hierarchical naming and flat naming schemes to a certain extent. It possibly requires more overhead than both hierarchical and flat naming schemes because of the fact that it has to work with both of them and have mechanisms that perform the functions that both naming schemes might possibly need. These mechanisms also need to be able to function seamlessly and switch between the various portions of the hybrid name. This is a challenge in and of itself, but the benefits are many. Hybrid names are self-certifying, can provide a degree of human readability, can assist in routing using both the hierarchical aggregation policies of hierarchical naming schemes and various routing techniques for flat naming schemes, and can be able to interwork with the various naming schemes already present, both flat and hierarchical.

2.5 Naming Schemes Currently Used in ICN Architectures

There are multiple implementations of ICN architectures today. They work very differently from each other. These architectures have been implemented and have had time to begin to grow. Work on these has been continuing for a good amount of time. Some of these architectures are NDN, DONA, PURSUIT, NetInf, CCNx, Mobility First, and Overlay ICN (O-ICN). Each of these architectures is designed with some performance indicators in mind. Depending upon the performance goal, their operations are also different and so are algorithms for naming, routing and caching policies, etc., and the way these architectures publish their content and provide access to the content also differs. Based upon that, the naming schemes followed by different architectures are also different, e.g., NDN advocates simple and human readable names and hence has used a hierarchical naming scheme for addressing the content, while SAIL and O-ICN architectures focus more on name security and readability than the simplicity of names; hence they adopt hybrid naming schemes. In the subsequent subsection, we discuss the naming scheme used in NDN architecture. Because the choice of naming schemes is strongly correlated with the design goals of the ICN architectures, readers are encouraged to refer to the respective ICN architectures for the reasons of naming used by a particular architecture.

2.5.1 Content Naming in NDN Architecture

In the NDN architecture, content is encouraged to have human readable hierarchical naming. Although the end result of the name depends on the name supplied by the application developer, there are some recommendations and guidelines that can help the application developer design his or her namespace. This would make the architecture more optimized since following well-established conventions will make it easier for the NDN architecture to handle such names.

A basic overview of how the NDN architecture fetches data is as follows. In the NDN architecture, there are two types of packets to help facilitate the query response model. This is done with the help of interest and response packets. In this model, there is a global announcement regarding the availability of data. If any user is interested in a data object, that user will send out an interest regarding the data object. This interest is then picked up by the various network resources and the routers/servers that have said packet reply with a response packet. The routers then perform longest prefix matching to determine how to send the packet to the source.

Coming to the naming scheme itself, it is a hierarchical name in a format similar to a file system naming scheme. Since the strings in the name can contain any character, the name can also contain any character but it is desirable that the names given be human readable and thus it is encouraged to use human readable strings in the name.

These names are encapsulated in a wire format as pure byte strings and are then encapsulated with the headers being similar to TLV-style headers. There are certain names that cannot be human readable due to various reasons like the presence of a signature, the presence of raw data in the name, some sort of versioning or marker requiring the use of special characters, etc. For these cases, there is a set of rules which governs the conversion from the human readable names to their NDN TLV counterparts. They are present in NDN TLV specification [37] and can be referred to for further information [3,4,8].

2.6 Naming and Addressing in Mobile Scenario

Mobility is a huge challenge that needs to be considered in any network today. The question of how to have a seamless transition from one connection to another without affecting impact is one that the legacy network has been solving for a long time now. However, unlike the various legacy networks which consider mobility as a feature that has been added after the design of the network, ICN architectures try to have mobility included as a part of the basic design.

However, unlike in the legacy network where mobility is usually handled by creating tunnels and having mobility anchors which can help create an end-to-end tunnel between them to allow for mobility within the scope of the anchors, in ICN there is no one method to implement mobility. Mobility is a feature that has been implemented in a unique manner addressing variable scopes by individual architectures.

The key objective of mobility in ICN is to ensure that packet delivery is not halted or impacted when an ICN device is mobile. The whole notion of connectivity being impacted during mobility is abstracted out from the perspective of the ICN application. This seems intuitive from certain perspectives. In ICN, the focus is on content delivery as opposed to content delivery between a pair. Thus, there is no focus on forming a tunnel between the mobile device and the servicing nodes. There are procedures that need to be done at the new joining point to validate and authenticate this device. It is desirable that these procedures be initiated before the connection here has broken so that from an application perspective, there is no difference and packets just have to transfer via a different route without loss.

From a naming perspective, this is important because the choice of naming scheme impacts the mobility. While concrete studies have not been performed comparing the naming schemes against one another with respect to mobility to the best of our knowledge, there is an implementation of mobility in most architectures to a certain level. This implies that to a certain extent, there is scope for mobility in all the naming schemes, based on the addition to the names and the mechanisms handling them. The only challenge is that a naming scheme should not impede mobility in any manner.

The implementation of ICN in mobile and cellular networks is a relatively newer topic with a lot of research scope. There are also works which have studied and proposed a possible ICN in LTE and ICN in 5G architectures, although these are at a very preliminary stage [36].

2.7 Security and Privacy Issues for Naming

Security is an important issue to consider in the world today. With the amount of critical data that is on the web, the data that can pass through it, and the growing technology enabling cyberattacks, security is a major concern. This is even more of a problematic scenario when one considers the fact that ICN is still in its nascent stage and has a lot of room to mature. This also implies that there are currently issues that cannot be designed for and will only improve as ICN has time to grow. This kind of growth requires time and effort, which from a security perspective is a challenge for the ICN paradigm.

Security in the context of ICN is a separate and vast research area in and of itself. There is a lot of discussion and research that is happening on the security and privacy space in ICN. In ICN, unlike the legacy IP network, security is considered in the design and architecture phase as opposed to as an afterthought. This helps the ICN discussion as this way, the ICN paradigm can evolve with security mechanisms built in that can be further enhanced upon. In this section, we will give a brief introduction to security, the kind of security that ICN is working on, and the kind of attacks that are specific to naming. We will not discuss each of the attacks in detail and will just provide an overview, but we will cover them in brief and point to resources that can give more information regarding the topic at hand.

So far security in ICN has been primary focussed on authentication. The data object needs to be authenticated and verified, it should be possible to authenticate and verify the subscriber, and the data object should be a valid answer to the query posted. These three items need to be addressed carefully from a basic authentication perspective.

It is important that a data object be identified uniquely and be able to certify itself. Is this the right data packet, is this packet complete and uncorrupted, and is this data packet actually bound to a real-world identity that a requester can use to verify the authenticity of the data—all of these are questions that can impact the security measures taken in ICN naming.

Naming attacks in ICN are generally the kind of attacks that can be classified into a sniffing attack or watchlist attack. Generally, in these attacks, the attacker looks to prevent information from reaching a particular host. This is done by using the flow of information and possible control over that flow. In ICN-based architectures, since the request and the object requested are visible to everyone, the attacker has access to such flows, or to be more precise, greater access to such information flows, and can thus gain more information on what the data being requested is and who is requesting it. This causes the requestor to be increasingly vulnerable.

Watchlist attacks are when attackers have a list of people that they target. They modify the requests sent by these users to better serve their purposes. They can possibly delete the requests and cause a denial of service, or they can modify the requests to request a corrupt packet which can then be sent to the requestor to stage other attacks and so on.

Sniffing-based attacks are attacks where the attacker keeps track of the data that is flowing through the network while keeping track of certain specific keywords. When the attacker finds data with that keyword, he or she can then modify it or take an action on it that may possibly impact service. Both of these are based on how the name and interest in a packet can communicate information.

The possible types of attacks that have been considered so far for names are censorship based, privacy based, and denial of service based. Each of these has its own impacts. There are methods that are being proposed to help reduce vulnerability against these attacks. The process of encrypted messages with a possible prior key exchange, self-certification of data to prove validity and integrity, etc. are all mechanisms that have been proposed.

As previously mentioned, security in ICN is a huge research area by itself. Those interested can look at the ICN research challenges RFC to get a basic idea and then move on to other literature to better understand the kind of problems that ICN security might face [11,21,22,27,32,35,36,38].

2.8 Future Research Directions

There is a lot of scope for research in naming. There are still a lot of challenges. Some of the ideas that can be considered are the way a naming scheme impacts factors like mobility and security. The way security implementations impact the performance of a naming scheme, the scalability of a naming scheme and factors to be considered in detail, whether naming schemes should make a difference in versioning of data object, how to coordinate between legacy infrastructure, existing naming schemes and proposed naming schemes, name resolution and routing optimized mechanisms for specific naming schemes, etc.

These problems are only with respect to naming and addressing and there are many more such problems in naming and addressing. Naming and addressing is one single area of research for ICN; routing, security, implementation in cellular, implementations for mobile, implementations for multimedia streaming, implementations for IoT, etc. are a few other topics mentioned. There are many more such topics that could possibly be worked upon.

ICN seems to be a promising alternative paradigm to the current Internet but it is still in its nascent stages as opposed to the Internet which is around two decades old if not more. Due to this, there are a lot of challenges that it faces that need to be resolved before it can be put into effect in actual large-scale networks.

2.9 Summary

In this chapter we discussed that naming is a challenging problem for the future of ICN. The huge rate of content generation further complicates the problem. There are a few ICN architecture existing in literature such as NDN, DONA, NetInf, etc. but these architectures do not capture the full scale of the Internet. A lot of work has been done in the naming scenario but most of the solutions are related to a particular architecture. It is generally observed that there is a deep coupling between the ICN architecture and the naming scheme used in the very architecture. In this work, the analysis is based on the relative strengths and weaknesses of the various naming schemes without being biased for or against any particular architecture.

Attribute-based naming does not seem likely to go forward w.r.t. naming in the ICN domain. Flat naming, hierarchical naming, and hybrid naming seem like the front-runners as suitable naming candidates for ICN networks. All three have relative advantages and disadvantages. The benefits obtained from a hierarchical naming scheme occur when there are a huge number of data objects, but during these times the hierarchical naming scheme also has a lot of drawbacks due to large and variable length names and significant overhead. Hierarchical names are human readable which is a desirable quality as well as easy to implement. Flat naming offers security and a unique identification but there is still research work required regarding issues of scale. Hybrid naming is not completely human readable but meets the human readability requirement in a satisfactory manner. Hybrid names tend to work well on most of the requirements but the naming scheme also has significant overhead resulting in implementation complexities as well as possible delays.

Table 2.1 contains the performance of each naming scheme theoretically on each attribute in a side-by-side manner, where "X" indicates that the naming scheme does not perform well theoretically while "Y" indicates that it performs well. Wherever the symbol "—"

Table 2.1   Tabular Summary of Naming Schemes on Various Factors

	ATTRIBUTE NAMING	FLAT NAMING	HIERARCHICAL NAMING	HYBRID NAMING
Uniqueness	X	Y	X	Y
Assistance in Routing	X	X	Y	Y
Scalability	—	X	Y	—
Security	X	Y	X	Y
Readability	Y	X	Y	—

occurs, it indicates that the performance of the naming scheme has not been completely ascertained.

Below there are a list of references and reading material that will be useful for people entering ICN naming research as a good starting line. One can also look at [20] to find ICN projects in work.

References

1. B. Ohlman, "From ID/Locator Split to ICN," *2015 12th Annual IEEE Consumer Communications and Networking Conference (CCNC)*, Las Vegas, NV, 2015, pp. 256–261.
2. H. Zhang et al., "Uniform Information with a Hybrid Naming (HN) Scheme," *IETF Internet draft draft-zhang-icnrg-hn-07, Work in Progress*, October 2017.
3. V. Jacobson et al., "Networking Named Content," *Proceedings of the 5th International Conference on Emerging Networking Experiments and Technologies*, New York: ACM, 2009.
4. L. Zhang et al., "Named Data Networking (NDN) Project," Technical Report, NDN-0001, October 2010. https://named-data.net/publications/techreports/tr001ndn-proj/.
5. T. Koponen et al., "A Data-Oriented (and beyond) Network Architecture," *ACM SIGCOMM Computer Communication Review*, New York: ACM, 2007.
6. C. Dannewitz, "NetInf: An Information-Centric Design for the Future Internet," *Proceedings of the 3rd GI/ITGKuVS Workshop on the Future Internet*, Munich, Germany, May 2009.
7. A. Carzaniga, M. Rutherford and A. Wolf, "A Routing Scheme for Content-Based Networking," *INFOCOM 2004. Twenty-Third Annual Joint Conference of the IEEE Computer and Communications Societies*, IEEE, 2004.
8. D. Perino and M. Varvello, "A Reality Check for Content Centric Networking," *Proceedings of the ACM SIGCOMM Workshop on Information Centric Networking*, Toronto, Canada, 2011.
9. B. Ahlgren, C. Dannewitz, C. Imbrenda, D. Kutscher and B. Ohlman, "A Survey of Information-Centric Networking." *IEEE Communications Magazine*, vol. 50, no. 7, pp. 26–36, 2012.
10. A. Ghodsi et al., "Naming in Content-Oriented Architectures," *Proceedings of the ACM SIGCOMM Work Shop. Information-Centric Networking*, Toronto, Canada, August 2011.
11. B. Hamdane, A. Serrhouchni, A. Fadlallah and S. G. E. Fatmi, "Named-Data Security Scheme for Named Data Networking," *2012 Third International Conference on The Network of the Future (NOF)*, Gammarth, Tunisia, 2012, pp. 1–6.
12. A. Ghodsi et al., *Information-Centric Networking: Seeing the Forest for the Trees*, Cambridge, MA: HotNets-X, 2011.

13. S. Farrell et al., "Naming Things with Hashes," *IETF Internet Draft Draft-Farrell-Decade-ni*, April 2012.
14. M. D'Ambrosio et al., "MDHT: A Hierarchical Name Resolution Service for Information-Centric Networks," *Proceedings ACM SIGCOMM Wksp. Information-Centric Networking*, Toronto, Canada: ACM, 2011.
15. N. L. M. van Adrichem and F. A. Kuipers, "Globally Accessible Names in Named Data Networking," *2013 IEEE Conference on Computer Communications Workshops (INFOCOM WKSHPS)*, Turin, Italy, 2013, pp. 345–350.
16. D. Kutscher et al., "The NetInf Protocol," *IETF Internet Dragt Draft-Kutscher-icnrg-netinf-proto-01*, August 2013.
17. Y. Wang et al., "Fast Name Lookup for Named Data Networking," *2014 IEEE 22nd International Symposium of Quality of Service (IWQoS)*, Hong Kong, 2014, pp. 198–207.
18. H. Yuan and P. Crowley, "Reliably Scalable Name Prefix Lookup," *Architectures for Networking and Communications Systems (ANCS), 2015 ACM/IEEE Symposium on*, Oakland, CA, 2015, pp. 111–121.
19. X. Zhang et al., "PID: A Generic Naming Schema for Information Centric Network," *IETF Internet Draft draft-zhang-icnrg-pid-namingscheme-03*, February 2014.
20. T. Edwall, "Scalable & Adaptive Internet Solutions (Sail)," 2011. https://sail-project.eu/.
21. C. Dannewitz et al., "Secure Naming for a Network of Information," *IEEE INFOCOM Computer Communications Workshops*, 2010.
22. X. Zhang et al., "Towards Name-Based Trust and Security for Content-Centric Network." *Proceedings of the IEEE ICNP*, 2011.
23. G. Xylomenos et al., "A Survey of Information-Centric Networking Research," *Communications Surveys & Tutorials, IEEE*, vol. 16, no. 2, pp. 1024–1049, 2014.
24. J. Pan, S. Paul and R. Jain, "A Survey of the Research on Future Internet Architectures," *Communications Magazine, IEEE*, vol. 49, no. 7, pp. 26–36, 2011.
25. G. Zhang, Y. Li and T. Lin, "Caching in Information Centric Networking: A Survey," *Computer Networks*, vol. 57, no. 16, pp. 3128–3141, 2013.
26. S. K. Fayazbakhsh et al., "Less Pain, Most of the Gain: Incrementally Deployable ICN," *ACM SIGCOMM Computer Communication Review*, vol. 43, no. 4, pp. 147–158, 2013.
27. X. Jiang, J. Bi, G. Nan and Z. Li, "A Survey on Information-Centric Networking: Rationales, Designs and Debates," *China Communications*, vol. 12, no. 7, pp. 1–12, 2015.
28. R. Ahmed et al., "αRoute: Routing on Names," *IEEE/ACM Transactions on Networking*, vol. 24, no. 5, pp. 3070–3083, 2016.
29. D. Zhang and H. Liu, "Routing and Name Resolution in Information-Centric Networks," *2013 22nd International Conference on Computer Communication and Networks (ICCCN)*, Nassau, Bahamas, 2013, pp. 1–7.

30. S. Shailendra, B. Panigrahi, H. K. Rath and A. Simha, "A Novel Overlay Architecture for Information Centric Networking," *2015 Twenty First National Conference on Communications (NCC)*, Mumbai, India, 2015, pp. 1–6.
31. S. Kim, Z. Duan and F. Sanchez, "Scalable Name-Based Inter-Domain Routing for Information-Centric Networks," *2015 IEEE 34th International Performance Computing and Communications Conference (IPCCC)*, Nanjing, China, 2015, pp. 1–8.
32. J. Hong, W. Chun and H. Jung, "A Flat Name Based Routing Scheme for Information-Centric Networking," *2015 17th International Conference on Advanced Communication Technology (ICACT)*, Seoul, South Korea, 2015, pp. 444–447.
33. J. Ren et al., "On the Deployment of Information-Centric Network: Programmability and Virtualization," *2015 International Conference on Computing, Networking and Communications (ICNC)*, Garden Grove, CA, 2015, pp. 690–694.
34. G. Carofiglio, G. Morabito, L. Muscariello, I. Solis and M. Varvello, "From Content Delivery Today to Information Centric Networking," *Computer Networks*, vol. 57, no. 2013, pp. 3116–3127, 2013.
35. K. Pentikousis et al., "Information-Centric Networking: Evaluation and Security Considerations." No. RFC 7945, 2016.
36. D. Kutscher et al. "Information-Centric Networking (ICN) Research Challenges." No. RFC 7927, 2016.
37. NDN tr-22 NDN Memo Naming Conventions. July 21, 2014. https://named-data.net/wp-content/uploads/2014/08/ndn-tr-22-ndn-memo-naming-conventions.pdf.
38. E. G. AbdAllah, H. S. Hassanein and M. Zulkernine, "A Survey of Security Attacks in Information-Centric Networking," *IEEE Communications Surveys & Tutorials*, vol. 17, no. 3, pp. 1441–1454, Third Quarter 2015.

3

INFORMATION-CENTRIC EXCHANGE MECHANISMS FOR IoT INTEROPERABLE DEPLOYMENT

DANIEL CORUJO, CARLOS GUIMARÃES, JOSÉ QUEVEDO, RUI FERREIRA AND RUI L. AGUIAR

Contents

3.1 Introduction

Information-Centric Networking (ICN) [1] research is often confronted in regards to its potential for out-performing and/or replacing the underlying IP fabric of today's Internet, in order to improve currently experienced operations, as well as to provide a more flexible platform to face the increasing set of challenges that arise from the integration of Information and Communication Technologies (ICT) in many different fields. The interest in ICN, as an enabler for a Future Internet, has been growing in the last few years, supported by industrial, academic and standardisation research efforts, illustrating the contribution that its novel design can provide in different deployments [2]. The ability to use a name to retrieve content, in a way that is location and access technology independent, and the capability for routers to take advantage of content caching, are only some of the intrinsic ICN aspects that may pave the way for a fundamental new Internet [3]. These aspects can further push new technological insights, by enhancing name-based mechanisms as seen in projects such as SeLF-ICN,[1] where names are used by network mechanisms to reference elements inside content itself.

However, the Internet, and everything around it, sees the IP protocol deeply rooted in its underlying mechanisms, rendering transition efforts very complex, not only from technical but also from business points of view. Despite evidenced issues with the current IP protocol and the increasing need for its evolution, as well as the manifestation of new mechanisms that have the potential to offer more features or improve performance only in specific scenarios, we are still a long

---

[1] https://atnog.av.it.pt/content/self-icn

way from proclaiming ICN as the *de facto* Future Internet candidate [4,5]. On the one hand, efforts to assert ICN deployment require a validation of its core mechanisms and potential contributions in a broader range of evaluation scenarios, one of which is the Internet of Things (IoT), a heterogeneous ecosystem of interconnected smart objects (e.g. SmartCities, SmartArgriculture and SmartHomes) [6]. On the other hand, both IoT and ICN face challenges related to the heterogeneous nature of their operational environments. ICN rollout transition strategies at later (more stable) stages still need to contemplate the existence of a mainly IP-based network [7] as well as different ICN instantiations. In IoT (even within IP), different deployment "silos" with specific solutions are not able to integrate horizontally with other IoT-cells, requiring consumer systems to support different approaches to obtain the desired heterogeneous content. As such, it becomes paramount for related research to clarify the capability of ICN for accommodating the necessary enhancements to its base architecture in order to: (a) better accommodate scenarios that, when subjected to ICN operation, evidence shortcomings when compared with their operation in current IP-based networks and (b) interoperate with existing networks, as well as among different ICN architectural instantiations.

This chapter addresses this dual problem with the objective of providing a concise view on the deployment capabilities of ICN where interoperation with other architectures is needed, specifically in complex scenarios whose stringent requirements demand enhancements to the base operations of ICN itself. Concretely, the chapter explores how an IoT deployment is impacted by the underlying mechanisms of ICN, evidencing not only the contributions that name-based operations can provide, but also identifying its shortcomings and necessary enhancements to ICN in order to avoid them. IoT represents one of the most challenging, demanding and heterogeneous areas of current ICT deployments. Supportive networking infrastructures started to have to deal with devices featuring widely different power, communication and computation capabilities, often simultaneously and through different access technologies featuring specific operational modes such as Wifi, Bluetooth, ZigBee or even LoRa[2] and SigFox,[3] and then connecting them to processing platforms. This complexity resulted in a very heterogeneous environment, where even the deployment of IP mechanisms (and resulting IP evolutions for IoT on their own) is limited or produces widely different results and solutions. As such, it provides an interesting opportunity for the assessment of ICN's contribution as a whole, in the provisioning of a simplified architecture towards IoT interaction, or by the individual operational components that compose ICN architectures [8,9]. Naturally, when ICN is subjected to an environment featuring stringent IoT requirements, optimizations to its base behavior become necessary. These questions will be addressed by evidencing, proposing and evaluating a set of key mechanisms that complement ICN in IoT environments fomenting heterogeneous connectivity. In targeting this problem an example of an interoperability framework is discussed, and it is able to convey information and content that need to traverse distinct network architectural domains. The potential contributions of enabling the discovery and the identification of content across different connected networking architectures and protocols are also addressed.

The chapter is organized as follows: after a brief introduction, Section 3.2 (Perspectives of ICN for IoT) explores the opportunities that ICN provides to IoT deployments, discussing existing approaches and standardization work, and evidencing how the generic traits of these Future Internet architectures contribute to such scenarios. Equally important, this section highlights potential impact points that the underlying operations of the ICN architecture can exert over key aspects of IoT, such as caching, information freshness, security and overhead, amongst others. Section 3.3 (IoT through the ICN Glass) starts by presenting the relevant works towards an efficient application of information-centric concepts for IoT solutions. Afterwards, it proposes a Publish-Subscribe protocol for Interest-based ICNs,

---

[2] LoRa: https://www.lora-alliance.org/
[3] SigFox: http://www.sigfox.com/

enabling efficient push-like communications in such architectures. The evaluation of this protocol showcases its suitability for IoT scenarios in which event occurrences are unpredictable (e.g. emergency notification). Section 3.4 (Exchange Mechanisms for Integrating the Heterogeneous Networking Landscape) provides a thorough look at interoperability aspects for generically integrating different networking architectures. This paramount subject recognizes the contribution potential of the novel mechanisms and utilization paradigm in solutions such as ICN, but acknowledges that ICN stands over an IP-based world. As such, in this section, a generic mechanism that enables different architectures to interoperate is proposed and evaluated. The solution tackles adaptation measures at different degrees, providing insights and measures for addressing, signaling and content conformity, according to the technological environment of the content source and requester. This insightful approach is showcased in terms of a typical Internet-based utilization, where, e.g., web browsing traffic needs to traverse different kinds of network architectures. Section 3.5 (Discovery and Mapping of IoT Content across Heterogeneous Architectures and Protocols) tackles the discovery, identification and interoperable addressing of content available in different connected networking architectures and protocols by consumers in other deployments. This focuses on the key point that one of the optimal means to introduce a potentially new (and better) protocol is to allow its native applications to reach for content existing in distinct architecture domains. For that, a solution that leverages cross-protocol discovery models is proposed, providing the means to adapt the description of existing inter-architecture content and the necessary inter-exchange signaling. Different deployment scenarios are analysed, exploring alternative interoperability opportunities. Section 3.6 (Information-centric Exchange Mechanisms for IoT Interoperable Deployment) then evolves into an integrated contribution, where the previous IoT enhancements for ICN, and the inter-architecture exchange mechanisms, are harmoniously combined. A scenario featuring different networking technologies operating as the vehicle for IoT information between distinct deployments is evaluated, comparing the impact of the proposal mechanisms over the current IP and the ICN. Section 3.7 (Management of Interoperable IoT Deployments) briefly highlights management aspects related to the presented solution. The chapter is then finalized with conclusions and the identification of outstanding issues in Section 3.8, and future research directions are provided in Section 3.9.

Despite the complex and heterogeneous nature of IoT as a whole, with its stringent operational requirements and procedures, it still addresses a specific scope, whereas ICN, as a network architecture, needs to be able to operate in all scenarios. However, the specificities and environment differences in IoT deployments are able to provide valuable insight, not only as a first step, but also as a provider of key enablers and extensions to the ICN base design, in order to more holistically support what is required of a full-fledged Internet mechanism, able to interoperate into other domains, independently of the deployment scenario.

3.2 Perspectives of ICN for IoT

ICN is a novel networking paradigm centered around content, unlike the current Internet architecture which follows a host-centric concept. This new approach enables new networking features such as naming, caching and securing content objects, which match some of the generic requirements generally associated to the Internet of Things. Notwithstanding, it is a recent and continuously evolving paradigm, initially conceived for content distribution scenarios. Exposing the ICN concepts to the heterogeneous applications and scenarios commonly associated with the IoT is expected to expand ICN mechanisms while empowering novel ways of IoT utilization. The current section explores the opportunities of using ICN for IoT deployments as well as the potential challenges that the underlying networking operations of ICN may impose over key aspects of IoT.

3.2.1 Opportunities

Information-centric approaches generally identify content with names, which are used for forwarding and routing throughout the network. Naming content at the network layer can be further leveraged for performing additional functions such as content filtering (e.g. /it/room136/sensor03/temperature is likely to be an IoT traffic involving information from a temperature sensor), in-network processing (e.g. a request like /it/room136/temperature/average will trigger multiple requests to get the information from all the temperature sensors in the room and will be responded with the average of all the received values) and content aggregation (e.g. /it/room136/temperature could have the aggregated content from all the temperature sensors in the referred room) which are relevant for an efficient IoT.

ICN, as its name suggests, centres its networking functions around content in contrast to the host-centric approach of the current Internet. By identifying content instead of hosts, ICN intrinsically supports content retrieval from any host holding a valid copy of the object, thus decoupling information consumers and producers, which leads to more efficient content distribution, ease in mobility and multihoming and better disruption tolerance [10]. These features are empowered when considered altogether with ICN in-network caching mechanisms and can be largely benefited in lossy wireless environments (statistically fragmented) involving low-powered devices (possibly resource-constrained and with long sleep cycles) commonly found in IoT scenarios [11]. Additionally, by retrieving cached content, and avoiding unnecessary transmissions, the scarce resources commonly associated with IoT-based devices (e.g. bandwidth, battery lifetime) could be saved, while reducing delays [12]. Moreover, proactive caching of content could be considered for achieving even better performance [13].

Along with the networking paradigm shift, ICN moves away from the channel security strategy into content-based security. As such, content may be retrieved from untrusted sources and through insecure channels. This fact gains relevance in IoT scenarios, because content is secured just once and then redistributed throughout the network, saving processing resources with respect to a channel-based security approach in which each consumer should establish an end-to-end secure channel, thus hindering the scalability of a secure IoT [9].

3.2.2 Challenges

The IoT is expected to be composed of a large number of heterogeneous devices. This heterogeneity is expected not only in terms of computational and communications capabilities, but also in the way information is structured and provided (e.g. content semantics, metric system). The resulting vertical silos have hindered the interoperability among different solutions and deployments, as well as the possibility of achieving more complex and truly global IoT environments. In IP networks, and other host-oriented networking approaches, these semantic issues should be, in general, addressed at the application layer. However, the fact that ICN networking functionalities are centred around named content makes semantic heterogeneity also a networking problem. Therefore, although the use of names may ease the IoT deployment and operation, there is a need to carefully tackle the heterogeneity in terms of naming and nomenclature to refer to IoT content. The use of semantic similarity has been studied as a possible way of addressing interoperability issues in ICN [14].

Besides naming semantic heterogeneity, the adoption of a naming scheme could be also conditioned by the target IoT scenarios. For example, while in some cases it is necessary to uniquely identify a piece of content (e.g. registering historical sensing data, addressing a particular actuator), in other cases it may be desirable to have mutable content (e.g. a live video surveillance stream). Consequently, conventions on how content should be addressed are required.

From an operational point of view, by having named content at the center of the networking functions, ICN is expected to raise other issues when considering the characteristics of IoT traffic. While IoT content is generally small, uniquely identifying it could require relatively long names leading not only to issues in terms of overhead, but also challenging constrained networks with low MTUs (e.g. ZigBee). Also, mechanisms for addressing content that does not yet exist (e.g. by describing the content using metadata) have to be implemented as it may be required in certain IoT scenarios (e.g. to force a sensor to perform a sensing operation and reply with the result). This last issue becomes even more critical when considering self-certifying namespaces in which the names depend on the data itself.

In general, ICN architectures follow a pull-based communication approach, either Request/Response or Publish/Subscribe. However, certain scenarios may require push-like communications (e.g. notification of an event). As such, mechanisms for emulating push-based communications should be defined. Additionally, although ICN networking operation is centered around content, in some cases it may also be desirable to address a particular host (e.g. sending an action to a particular actuator, management-related communications).

Despite the fact that different ICN solutions intrinsically implement security mechanisms, achieving security levels in compliance with the requirements of the IoT calls for further development. Moreover, the cryptographic algorithms considered by ICN security mechanisms could be too demanding for constrained devices, thus requiring different algorithms (e.g. elliptic curve cryptography) or even securing strategies (e.g. delegation of cryptographic functions). Additionally, some security dimensions are more complex to address and have to be redefined from the way they are currently achieved. For example, in ICN it should be assumed that any consumer can access any piece of content (e.g. by retrieving it from a cache); consequently, controlling the access to the information should be done in a different way (e.g. by obfuscating the content itself) [15].

In-network caching is recognized to be one of the prime contributions of the information-centric paradigm. However, some IoT scenarios may require close to real-time communications, which not only diminishes the advantages that caching systems may bring to the overall performance, but may also lead to the retrieval of obsolete information. Moreover, in ICN systems involving distributed caching, producers have reduced control over the content replication and therefore it is technically challenging to find all copies of a given content and mark them as staled. Although some mechanisms have been considered by the different ICN solutions to address this issue (e.g. use of a freshness parameter in NDN Data packets which defines the time a given piece of content can be cached by a given node before it becomes stale), they are still insufficient for a full control of the information freshness (e.g. because of the stackable nature of the freshness parameter, i.e., a Data could be retrieved from a given node just before it is marked as stale and therefore be cached again for the same time and then it can happen all over again repeatedly).

ICN caching may be done as the content is transmitted towards the consumer (i.e. on-path caching) or by intentionally placing the content in a specific network element (i.e. off-path caching) [1]. While the former caching solution is simpler and requires presumably no signaling, the second approach requires additional mechanisms to be enforced for managing the caching process, thus requiring additional overhead in terms of communication and processing. Moreover, achieving optimal content retrieval (e.g. identifying the closest off-path cached copy of the content) remains a challenge. In terms of security, by having caching at the network layer, ICN must carefully address attacks targeting the caching mechanisms (e.g. pollution and polling attacks).

3.3 IoT through the ICN Glass

In this section, the relevant efforts for the application of ICN principles in IoT environments are highlighted and a publish/subscribe approach on top of the Named Data Networking (NDN) solution [16] is presented along with its advantages when there is high uncertainty about the information generation process (e.g. notification scenarios).

3.3.1 *Contributions towards Efficient IoT Operation*

Over the last few years, we have been witnessing continuous development of the concepts associated with the ICN paradigm, as well as its target scenarios. Moreover, the ICN paradigm has been commonly associated with the IoT, with numerous research highlighting the convenience of applying information-centric concepts to address IoT's common challenges, contributing not only to the realization of more complex IoT scenarios but also to the evolution of ICN concepts and solutions. Notoriously, the Information-Centric Networking Research Group (ICNRG)[4] of the Internet Research Task Force (IRTF) has identified IoT as a baseline scenario where the use of ICN could bring significant advantages as compared to existing Internet protocols [2].

The literature [8,9,17] summarizes, from a theoretical perspective, the main benefits, challenges and guidelines for the proper application of information-centric concepts in IoT scenarios. Other works have targeted specific challenges and/or applications involving ICN concepts in IoT scenarios. For example, the use of long-term Interests for enabling push-like communications [18]; lightweight security alternatives for constrained devices [19]; authenticated Interests and encryption-based

---

[4] https://irtf.org/icnrg

access control for secure actuation [20] and sensing [21]; enabling of multi-source data retrieval [22]; management of IoT deployments over ICN [23]; impact of caching in energy and bandwidth efficiency [12]; information freshness [24]; and security management [15]; enabling discovery in IoT environments for local domain [25] and interoperable [YY] scenarios. Finally, an experimental analysis of the impact of applying ICN in IoT environments is presented in [26], demonstrating the feasibility of using ICN on top of constrained devices and showcasing the advantages over approaches based on 6LoWPAN/IPv6/RPL.

3.3.2 Named Publish Subscribe Networking (NPSN)

Inspired by the operational principles of the MQTT protocol [27], and by its wide adoption in diverse IoT scenarios and use cases, Named Publish Subscribe Networking (NPSN) protocol was developed in [YY] for enabling a Publish/Subscribe communication on top of NDN. The goal behind this proposal was to overcome NDN limitations derived from the pull nature associated with this network protocol. In particular, the NPSN protocol demonstrated to have better performance in terms of networking overhead and information freshness, with respect to the equivalent request/response approach in environments where the time associated with the generation of information is highly variable. In this section, we briefly introduce the fundamentals of NDN and NPSN. A novel utilization mode in which an NPSN entity can act as a standalone publisher (i.e., simultaneously providing the functionalities of the original rendezvous and publisher entities) is proposed.

3.3.2.1 Named Data Networking (NDN)   Named Data Networking (NDN) is a research project[5] from the *National Science Foundation* which focuses on a Future Internet Architecture. Based on the original code of CCNx,[6] NDN produced a derivative architecture and open-source implementation, sharing the key architectural principles of ICN. In that sense, NDN enables a receiver-driven communication based on Interest (i.e. a packet used by consumers to request content) and Data (i.e. a packet used by any entity in the network to

---

[5] https://named-data.net/project/
[6] https://blogs.parc.com/ccnx/

reply back with the requested content when it is available in its cache). The exchange of these packets is driven by two main aspects. The first one refers to the included name of the content, which allows it to be uniquely identified. The second is the set of data structures existing in ICN-enabled routers, namely the Pending Interest Table (PIT), which stores requests for content that have not been replied to yet; the Forwarding Information Base (FIB), which maps names to interfaces; and the Consent Store (CS), which acts as a temporary cache for content that is replied back to requests. This unique combination allows the requested content to be decoupled from where it was obtained, supported by security measures applied directly to the content instead of the communication channel. The NDN protocol is based on a request-response communication model that decouples the consumer and producer of the information. Moreover, consumers do not require a network address to retrieve content, and exchange messages do not contain consumer identification information. As such, this operation principles hinder the utilization of NDN in scenarios requiring the spontaneous push of content (e.g., emergency notification). A possible way of supporting such scenarios in NDN is to continuously inquire the network about the existence of new information (i.e., polling approach). However, in IoT scenarios where devices are not always available (e.g., because of sleep cycles) and communications should be minimized in order to save scarce resources (e.g., energy), the practicability of a polling approach may be limited. These limitations are more impactful as the information generation is more unpredictable.

3.3.2.2 NPSN Protocol Overview   Inspired in the operation of the MQTT protocol, NPSN emerges to account for those scenarios where NDN request-response approach is suboptimal, by providing a Publish/Subscribe communication model on top of NDN.

NPSN works under the following principles:

- The consumer has a routable prefix which will be used for listening requests from the producer.
- The producer has to be informed about the consumers' prefix and willingness to receive further information (i.e. connect and subscribe operations).

- The producer makes a request to subscribed consumers to let them know about the existence of new information (i.e. publish operation).
- Finally, consumers make a request to obtain the newly generated information.

In this work, we propose to add a Standalone publisher operation mode as an alternative to the original Rendezvous-based mode. While the original NPSN operation mode considers a dedicated entity responsible for matching subscriptions and publications (i.e., an operation similar to that of a MQTT broker), the proposed extensions assume that the publisher will manage subscriptions by itself. The signaling associated with these two modes of operations are presented in Figures 3.1 and 3.2, respectively. The signalling for the rendezvous-based operation mode is described in detail in [YY].

In the standalone operation mode, the subscriber sends a *connect* Interest (1) to let the publisher know the prefix at which it can be reached. In response, it receives a Data (2) with the session *id* generated for the subscriber. The reasons for using a *connect* message and generating an *id* for each client are twofold (i) **avoiding unnecessary overhead**: each time a subscriber sends a request it has to identify itself, thus sending the full client prefix (potentially large) could incur in additional protocol overhead, and (ii) **ease the parsing of the request**: the client prefix could contain a variable amount of components, thus requiring separators for indicating the end of the client prefix, whereas using an *id* leads to a fixed length of one component. Afterwards, the subscriber sends an Interest (3) using the provided session *id* and including the topic(s) to be subscribed. The publisher acknowledges the topic subscription with an empty Data (4) (this can be extended to provide more information about the subscription process for example, subscription lifetime). When the publisher has new data for a topic, it sends a *publish* Interest (5) to all subscribers for that specific topic. Subscribers then acknowledge using an empty Data (6) and send a *request* Interest (7) to obtain the new information. This information is finally provided in Data (8) by the publisher. The subscriber retains the subscription as long as it desires to receive publications (it can subscribe more than one topic as well). When a subscriber no longer wants to remain subscribed, the subscriber can send an *unsubscribe*

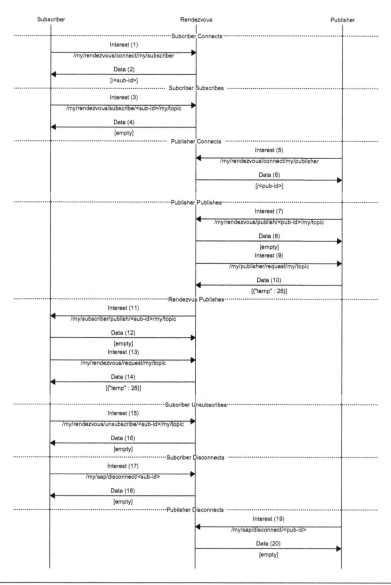

Figure 3.1   NPSN signaling for rendezvous-based operation as in [YY].

Interest (9) which is also acknowledged with an empty Data (10). Finally, if the subscriber wants to finalize the communication it sends a *disconnect* Interest (11) which is also acknowledged with an empty Data (12).

As in the original mode, the *request* Interest does not identify the subscriber and can therefore be requested by any entity of the network

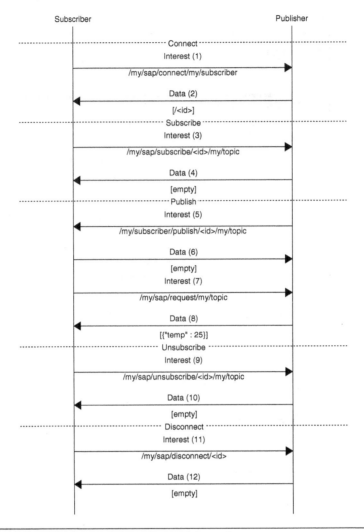

Figure 3.2   NPSN signaling for standalone publish operation.

(even if not connected or using the regular NDN instantiation). This ensures that the generic advantages of NDN are maintained (e.g., those associated with caching mechanisms) as well as compatibility with NDN.

3.3.2.3 Discussion   The NPSN protocol, as showcased in [YY], brings the benefits commonly associated with publish-subscribe approaches into NDN, thus improving the native request-response communication model in scenarios where the information generation is unpredictable. Notwithstanding, if the consumer knows beforehand when new information will be available (e.g., new information is generated periodically) following polling approaches could be more efficient (i.e., less packets are required to retrieve the information).

As such, it is important to enable the use of NPSN in scenarios where an NPSN rendezvous may not always be available (e.g., in ad hoc IoT environments). In such scenarios, the subscriber needs to be able to directly interact with the publisher of the information, requiring the latter to assume the mantle of functionalities previously provided by the NPSN rendezvous. In this chapter, a new operation mode accounting for these new requirements is proposed as an extension to the NPSN protocol.

However, a rendezvous-based mode is advisable when publishing nodes are constrained. The use of a rendezvous reduces the requirements imposed over the publishing nodes, not only by reducing the amount of processing requirements associated with the management of subscriptions but more importantly by reducing the interactions with the producer itself (i.e., the publisher communicates with one or more rendezvous entities instead of the potential large amount of subscribers). Additionally, it allows the rendezvous to perform more operations such as data processing and aggregation (e.g. average temperature in a room). In terms of security, the use of a rendezvous is also recommended, as simpler security mechanisms may be implemented in the publisher-rendezvous interface to account for constrained devices. The way in which security is managed is outside the scope of the current work, but readers can glance at security trust schemes for NDN in [28] and particularly in the context of IoT in [15]. Consequently, hybrid solutions where more powerful publishers act as a rendezvous point for more constrained publisher may be also considered.

3.4 Exchange Mechanisms for Integrating the Heterogeneous Networking Landscape

The adoption of novel network architectures, independent of their contributions, may be a slow and costly process, requiring architectural changes in existing infrastructures ranging from the network up to the application layers. To spread deployment costs over time, these novel architectures might be incrementally deployed through globally connected islands in parallel with the existing network landscape. However, it may lead to the creation of resource silos, where resources are restricted to their environment alone, not being able to be accessed by entities from different architectures. Mechanisms to enable the interoperation between different architectures are required to allow a holistic communication between entities independent of their access network architecture.

Research into the interoperability mechanisms for Future Internet architectures started not only as an enabler for the coexistence of different architectures but also for their incremental deployment in the existing networking ecosystem. In [29] a framework that aims to facilitate the deployment of different ICN architectures is proposed, also enabling the interoperation among the various ICN instances using tunneling or translation interoperation schemes. Other works follow a gateway or proxy-based approach to allow the traffic between HTTP endpoints to cross a CCN network [30]; use ICN as an underlay for IP-based services [7]; or carry TCP traffic between TCP/IP endpoints over an ICN network [31].

3.4.1 Challenges

These novel network architectures, aiming to efficiently solve different requirements of the current and Future Internet, centered their design on different network elements, resulting in the definition of a variety of network environments with distinct characteristics.

Next, we briefly describe potential aspects that can hinder the inter-operation between different architectures:

- **Different design orientations:** By centering their network layer on different elements, it may change how the network and applications are designed, developed and deployed due to different operation principles between architectures.
- **Different communication paradigms:** By supporting different ways of obtaining information, a proper information exchange between entities supporting different network architectures may not be possible.

- **Different addressing schemes:** By defining different addressing schemes, which can have different meanings, semantics and encodings can be scattered among different layers, and a direct address translation/inference between architectures may not be possible to achieve.
- **Different network protocols:** By defining different network protocols, equipment from a given network architecture may not recognize a received message belonging to other architectures, discarding the message and failing to deliver it to its destination.

3.4.2 A Mechanism for Network Interoperability

In this section we describe the proposed solution to integrate the heterogeneous networking landscape, enabling the interoperability between different network architectures.

3.4.2.1 Generic Architecture   Motivated by the need to provide interoperability between different architectures that coexist in a heterogeneous network landscape, we propose an intermediary entity, called Future Internet eXchange Point (FIXP), that enables entities deployed on different architectures to communicate with one another. In Figure 3.3, a supporting infrastructure composed of several FIXPs is depicted. In this example, the network resource NR, physically deployed in the network architecture FI_2, can be accessed from clients in the network architecture FI_1 (i.e. C_1) but not from clients in the network architecture FI_3 (i.e. C_3). The reason is that the $FIXP$ made the resource NR available in FI_1 through a virtual representation (i.e. NR'), enabling clients in that architecture to access the resource through addresses/mechanisms compatible with the protocols supported by that network architecture, whose occurrence is not verified in the FI_3.

The FIXP is an intermediary entity responsible for converting messages from a given network architecture into messages compatible with a different architecture. In doing so, it emulates the communication endpoints, behaving as the destination and the source of exchanged messages. For example, in Figure 3.3, both the client (i.e. C_1) and the network resource (i.e. NR) communicate with the FIXP, which is seen by these entities as an ordinary network resource (i.e. NR') and client (i.e. $C1'$), respectively.

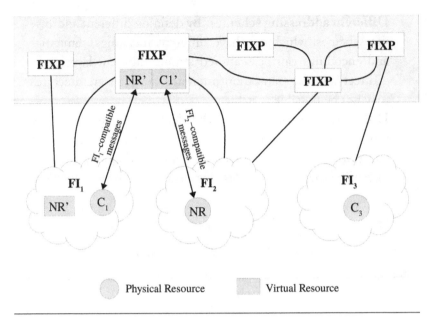

Figure 3.3 Solution overview.

In doing so, the interoperability procedure becomes transparent for both endpoints and network entities on each network architecture. Thus, it allows backwards compatibility with existing applications, services and mechanisms, as well as enables the development, implementation and evolution of each network architecture without imposing any restrictions in terms of interoperability.

Whenever a message is received by the FIXP, three main operations are performed: (i) address translation; (ii) signaling adaptation; and (iii) content adaptation.

3.4.2.2 Address Translation The address of the resource to be accessed may not be directly identified in the network layer, but scattered amongst multiple layers. For example, the network layer address may consist of a hint on how to reach the resource (e.g. the IP address of the server hosting the resource), being the resource itself addressed in the upper layers. As such, Uniform Resource Identifiers (URIs),[7] comprehending all the required information to identify the resource, are used in the proposed solution to uniquely identify the resource to be accessed.

---

[7] RFC 3986: https://tools.ietf.org/html/rfc3986

Before a given resource can be accessed beyond the network architecture where it is deployed, a URI pointing to the *real* resources, compatible with the protocols/mechanisms supported by each network architecture, needs to be made available by FIXPs. Thus, a resource is identified by a single original URI (i.e. the URI on the network architecture where it is physically deployed) and by multiple foreign URIs (i.e. the URIs that define how resources are accessed on the remaining network architectures).

The foreign URIs are automatically generated by a FIXP for all known network architectures as existing resources are being discovered, which are then synchronized among the remaining FIXPs. Several strategies to generate the foreign URIs could be used, such as:

- **Concatenation:** Additional information is appended to the original URI in order to comply with the destination protocol specificities or forwarding requirements.
- **Hash-based:** The foreign URI contains a hash of the original URI of the resource to be accessed.

Regardless on how foreign URIs are generated, they must meet two base properties: (i) the request for the URI is forwarded towards a FIXP and (ii) the resource is uniquely identified on the original network architecture.

Finally, foreign URIs define that the access to the resources is made via existing protocols on each architecture, allowing applications, services and network mechanisms to perceive the resources as being deployed in the same network architecture.

3.4.2.3 Signaling Adaptation The first step upon the reception of a new message by the FIXP is to identify the protocols that compose the message, so it can be forwarded to the correspondent protocol stack. If the message is identified as belonging to the transferal of information, it is marked as requiring conversion and, therefore, the FIXP computes its URI and extracts the payload and relevant metadata from the received message, which will then be used to convert the message to the destination architecture. Otherwise, the message is handled internally according to the corresponding protocol stack.

The FIXP looks up the corresponding URI to use on the destination architecture. Using this URI, the FIXP creates the destination protocol message, to which it adds the payload extracted from the received message and applies the metadata. The created message is then sent to the destination network architecture.

Since the FIXP can create the correspondent protocol message identified by an URI, to which it applies the payload and a set of metadata using protocol specific mechanisms, conversion algorithms may not be required for each pair of protocols.

3.4.2.4 Content Adaptation Besides converting the signaling between architectures, message payload may also need to be adapted in order to be compliant with the destination architecture. For example, assuming that *NR* in Figure 3.3 is the homepage of a given website, URIs contained in the web page (i.e. links to other resources) are bonded to the protocols supported by the network architecture where the web page is deployed. A web browser in C_1 is able to transparently fetch the web page, enabled by the proposed FIXP. However, if the web page content is not adapted to be in compliance with the network architecture of C_1, further requests by the web browser will be discarded. Among the reasons are (i) the protocol required to request a given URI is not supported by the browser or by the network stacks in the client and/or (ii) the network architecture does not support the required protocol and discards the message.

As such, during the signaling conversion and for specific types of content, the FIXP may inspect the payload of the messages to be converted in order to find exiting resource identifiers (i.e. URIs), which may be also converted to be in compliance with the destination network architecture if required.

3.4.3 Use Cases for Interoperability across NDN,
*   PURSUIT and IP Architectures*

In this section we describe a set of use cases, exposing the flexibility and capabilities of the FIXP to enable interoperability between different network architectures (considering their public specifications at the time of writing of this work). Specifically, our use cases include the interoperation between HTTP over TCP/IP (referred to as IP[HTTP]) and NDN architectures (i.e. oriented on different design paradigms), as well as between PURSUIT and NDN (i.e. based on different communication models). In both cases, the scenario consists of fetching an HTML page (Source Code 1) deployed in the NDN architecture, including other resources that composes it.

SOURCE CODE 1:   HTML SOURCE CODE

```
<html>
<body>
<img src="logo.png">
<p>Hello World!</p>
<a href="ndn:/atnog.av.it.pt">link</a>
<body>
<html>
```

3.4.3.1 Interoperability between IP[HTTP] and NDN Network Architectures Figure 3.4 presents the signaling exchanged between the different entities, when considering an interoperability scenario between the IP[HTTP] and NDN architectures, allowing HTTP consumers over the current Internet to transparently access resources deployed in the NDN architecture.

The consumer already knows the IP[HTTP] URI of the web page it wishes to receive. Firstly, it performs a DNS query which indicates the address of a FIXP that is seen in the IP network ecosystem as the *real* provider on behalf of the original NDN provider. Thus, it establishes a TCP connection with the FIXP (steps 1 to 3, in Figure 3.4), after which it sends the HTTP message requesting the desired resource (step 4). As shown in the signaling, the TCP control messages are not propagated to the NDN network, being handled internally by the FIXP, because they are related with the session established between the consumer and the FIXP. On the other hand, the *HTTP Get* message is identified as belonging to the transferal of information, being converted into an *NDN Interest* message to be sent through the NDN network (step 5b). The *NDN Interest* message is sent towards its original provider based on the existent mechanisms on the NDN network, without any awareness regarding

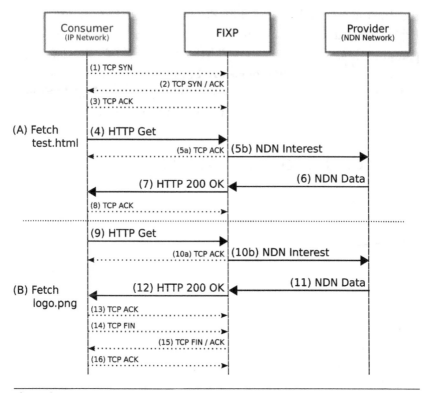

Figure 3.4 Interoperability signaling between IP[HTTP] and NDN.

the conversion process. The provider, or any cache in the path with the requested content, can reply with a *NDN Data message* towards the FIXP (step 6). Upon reception of this message, the FIXP extracts its content, which it uses to create the HTTP response message (step 7). During this step the FIXP inspects the content (in this case the HTML) looking for identifiers of other resources (i.e. URIs), replacing them if required in the content itself. If the URI mappings are not known, the FIXP generates the foreign URIs to be used in the IP architecture. Thus, content can maintain a full compatibility in the foreign architecture, allowing a transparent conversion procedure for the end-points and without the intervention of the resource provider. Subsequent resources that compose the HTML page are acquired using the same procedure described above (steps 9–13), after which the TCP connection between the consumer and the FIXP is terminated (steps 14–16).

3.4.3.2 Interoperability between PURSUIT and NDN Architectures The signaling supporting the interoperability scenario between PURSUIT and NDN architectures is presented in Figure 3.5. Since, in the current work, the PURSUIT URIs are generated without previous knowledge of the actual content of the resources being addressed, the generated URIs are not self-certified.

The FIXP, upon discovering the existence of new resources (i.e. after generating the mappings to be used in the PURSUIT network architecture), publishes them in the *Rendezvous Node* on behalf of the NDN provider, being seen in the PURSUIT network as the *real* provider. The PURSUIT consumer starts by subscribing the desired content towards the *Rendezvous Node* (step 1 in Figure 3.5). The *Rendezvous Node* requests the *Topology Manager* to calculate the forwarding identifier and sends it to the publisher (i.e. the FIXP) (step 2). The subscription is then converted by the FIXP into an *NDN Interest* message and sent towards the original content provider (step 3). As in the previous example, the NDN network is not aware that a conversion occurred and forwards the message based on its internal mechanisms. Any entity containing the content (i.e. the provider or caches) can reply with an *NDN Data* message towards the FIXP (step 4). The FIXP extracts the content, converting any identifier in the content itself to match the PURSUIT architecture,

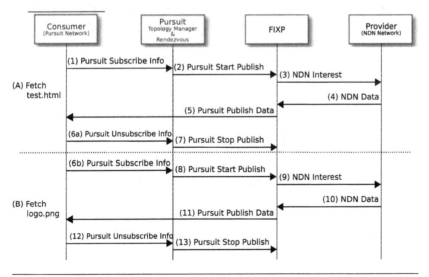

Figure 3.5   Interoperability signaling between PURSUIT and NDN.

and encapsulates it in a *PURSUIT Publish Data* message (step 5). Additionally, whenever the FIXP discovers a new PURSUIT URI (e.g. resulted from content conversion), it advertises itself to the Rendezvous as being the provider of the referred resource. Upon the reception of the HTML content, the client unsubscribes it with the *Rendezvous* (step 6a), and subsequently with the FIXP (step 7), and requests subsequent resources using the same procedure described above (steps 6b and 8–13).

3.4.4 Preliminary Results

In this section we present preliminary results for the proposed framework, focusing on the use cases described in the previous section. We aim not only to demonstrate the feasibility of our framework in interoperating different network architectures, but also to provide a first assessment of the cost of such an operation, to serve as the baseline for improvements in future research.

3.4.4.1 Performance Evaluation   The evaluation scenario (Figure 3.6) is composed of a set of clients (C1, C2 and C3), each supporting a single network architecture stack (NDN, IP and PURSUIT, respectively), and two NDN content providers (P1 and P2) accessible only by NDN procedures. The content provider P1 is connected to the FIXP via Router 2 (supporting only the NDN stack), while the remaining entities are connected to the FIXP via Router 1 (supporting IP, NDN and PURSUIT stacks).

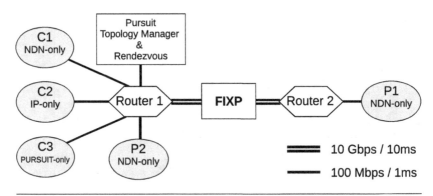

Figure 3.6   Evaluation scenario.

This scenario was implemented in a simulation environment using ns-3,[8] enhanced with NDN (ndnSIM 2.0 [32,33]) and PURSUIT (Blackadder[9]) network stacks, running on a virtual machine (with two 3.33 GHz CPU cores and 2 GB of RAM) hosted in an OpenStack Platform.

Our proposed framework enabled both IP[HTTP] and PURSUIT clients to fetch the HTML page (HTML and PNG files with size 119 and 6132 bytes, respectively) deployed in P1 and P2, transparently to both end-user terminals and network entities on each network architecture. Table 3.1 depicts the signaling size regarding each network architecture (for a single link) during the interoperability process, with the impact on the total exchanged information when retrieving the content from P1 and P2, shown in Figure 3.7. Figure 3.8 presents the cost of enabling interoperation in terms of fetching time. Since the NDN-NDN approach does not require message conversion, its values are presented as a reference to better evaluate the impact of the IP[HTTP]-NDN and PURSUIT-NDN approaches. Also, an IP-IP approach (referred to as "IP reference"), where all entities are IP-capable (including both providers P1 and P2), was also measured so it could serve as a reference value of the current Internet approach. Figure 3.9 details the delay introduced by the FIXP due to message conversion.

Compared with the NDN-only approach, the amount of signaling exchanged in the foreign network architecture for both IP[HTTP] and PURSUIT approaches is higher (going from 4 messages

Table 3.1   Signaling Size Regarding Each Network Architecture (For a Single Link)

	TOTAL SIZE W/ CONTENT (BYTES)	SIGNALING SIZE W/O CONTENT (BYTES)	EXCHANGED MESSAGES
C1 NDN	NDN: 7078	NDN: 827	NDN: 4
C2 IP[HTTP]	IP: 7345 NDN: 7078	IP: 1093 NDN: 827	IP: 19 NDN: 4
C3 PURSUIT	PURSUIT: 9355 NDN: 7078	PURSUIT: 2929 NDN: 827	PURSUIT: 10 NDN: 4
IP Reference	IP: 7345	IP: 1093	IP: 19

---

[8]   ns-3: https://www.nsnam.org/

[9]   Blackadder: http://www.fp7-pursuit.eu/

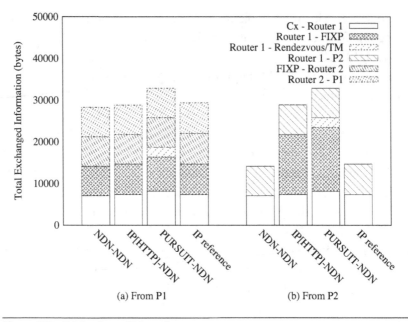

Figure 3.7   Total exchanged information.

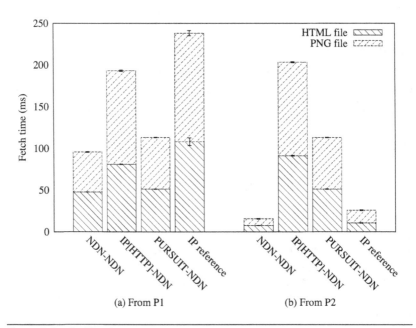

Figure 3.8   Fetching time.

to 10 and 19 messages, regarding PURSUIT and IP[HTTP], respectively). If in the IP[HTTP] approach the TCP control signaling and the configured IP MTU (due to which 5 TCP Data messages were required to transfer the whole PNG file) are responsible for the increase of exchanged messages, in the PURSUIT approach it is mainly due to the communication with *Rendezvous* required to (un)subscribe the different resources. This increase of exchanged messages resulted in an additional total size of 3.8% and 32.2%, respectively, for IP[HTTP] and PURSUIT. If the size of the content is not considered, the protocols in the PURSUIT and IP architectures saw their exchanged signaling increased, respectively, to 2102 and 166 bytes. In the original architecture neither the number of messages nor their size changed, since the same NDN messages were exchanged in all approaches.

When retrieving the contents from the provider *P1* (Figure 3.8a), IP[HTTP] and PURSUIT clients took longer to fetch the contents than when using an NDN-only approach (representing an increase of 99% and 17%, respectively). The main reason for the higher fetching times is the delay introduced by the internal mechanisms of both IP[HTTP] (i.e. (i) TCP control signaling and (ii) configured IP MTU, due to which 5 TCP Data messages were required to retrieve the PNG file) and PURSUIT networks (i.e. communication with the *Rendezvous*). Nevertheless, all the previous approaches proved to have lower fetching times as compared to the IP reference (i.e. IP-IP approach). This behavior is associated with the delay introduced by the TCP control signaling raised because of the higher RTT between the TCP endpoints (i.e. the client and the provider) in the IP reference approach. In terms of total exchanged information (Figure 3.7a), the higher amount of information exchanged to retrieve the HTML page using the IP[HTTP]-NDN and PURSUIT-NDN approaches, when compared with the NDN-NDN approach, is mainly due to the signaling overhead introduced by both IP[HTTP] and PURSUIT protocols.

However, when considering a nearest provider *P2* (Figure 3.8b), the gap in terms of fetching time between IP[HTTP]-NDN and PURSUIT-NDN approaches when compared with the NDN-NDN approach is even higher, due to the anchoring-related issues (i.e. longer and/or sub-optimal forwarding paths) of the proposed framework. This is also reflected in terms of total exchanged information (Figure 3.7b), where additional information is exchanged through the link between *Router 1* and *FIXP*. While the messages in the NDN-only approach only go through *Router1* between the client and the provider *P2* (i.e. *C1↔Router1↔P2*), in IP[HTTP]-NDN and PURSUIT-NDN approaches messages need to be forwarded to the FIXP in order to be converted to the destination network architecture (i.e. *C2/C3↔Router1↔FIXP↔Router1↔P2*).

Finally, in relation to the selected approaches, the delay introduced by the FIXP (Figure 3.9) in the communication between the client and the provider is approximately 1.3% and 3% for IP[HTTP] and PURSUIT, respectively. Converting IP[HTTP] or PURSUIT request messages to be compatible with the NDN architecture (i.e. to *NDN Interest* messages) is less time-consuming than the reverse process (i.e. from an *NDN Data* messages), mainly due to integrity check mechanisms.

This is more clear in the IP[HTTP] use case since the conversion from IP[HTTP] to an *NDN Interest* message is straightforward (i.e. the integrity of received IP[HTTP] messages is not verified), while in the other way the integrity of the contents contained in the *NDN Data* message are verified before converting the message to an IP[HTTP] response. This gap is smaller in the PURSUIT case

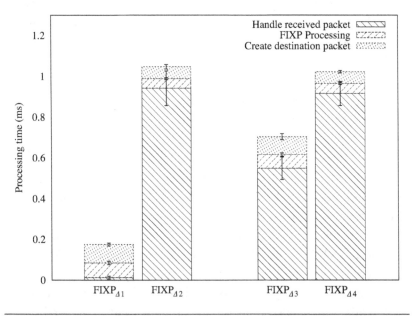

Figure 3.9 Detailed FIXP delay. (FIXP$_{\Delta1}$: Conversion from IP[HTTP] Request to NDN Interest; FIXP$_{\Delta2}$: Conversion from NDN Data to IP[HTTP] Response; FIXP$_{\Delta3}$: Conversion from PURSUIT Start_Publish to NDN Interest; FIXP$_{\Delta4}$: Conversion from NDN Data to PURSUIT Publish_Data).

because, in contrast to the previous case, the FIXP also verifies the integrity of the PURSUIT message before converting it to an *NDN Interest* message.

3.4.4.2 Major Lessons   The previous evaluation validates the practicality of the proposed solution to enable the interoperation between network architectures based on different design choices. In addition, this enablement should be considered as a solution that does not imply any modification of existing network entities (including user devices and applications). Although our solution introduces delay in the communication between endpoints due to the conversion procedure, in some conditions it was able to achieve better performance in terms of fetching time than an IP-only approach, as highlighted in the previous results.

Anchoring-related issues (Figures 3.7b and 3.8b) and sub-optimal support of other architectures features are among the problems that hinder the performance of the proposed solution. While the former depends on the deployment location of the FIXPs, which affects the total exchanged information and fetching times, the latter is caused by the workarounds used to support features from other architectures which may affect the number of messages (and consequently the information) exchanged (Table 3.1).

Moreover, a message can be converted and forwarded to the destination architecture only after being completely received by the FIXP, so it can apply any architecture-specific processing (e.g. integrity checks) and adapt the content to the destination architecture if required. Both operations may introduce an additional delay in the communication between endpoints. Nevertheless, not all content requires conversion.

3.5 Discovering and Mapping IoT Content across Heterogeneous Architectures and Protocols

The Internet of Things brings forth a vision that implies a myriad of device types communicating across multiple protocols. This vision implies the existence of mapping systems able to bridge the gap between protocols and their running environments.

This section addresses new possibilities towards a unified cross-protocol service discovery identification framework, supporting the deployment of this kind of IoT scenario. This approach facilitates deployment for services and mapping gateways, allowing services to embed service identification information over common discovery protocols, which clients can use to perform advanced discovery functions without complex IoT management systems.

3.5.1 Deployment Scenarios and Challenges

Device and service discovery is a fundamental function for any network, regardless of size or operational purpose. Over the years we have adopted a large number of protocols for the purpose of discovery in the network, which led to fragmentation of implementations and services concerning which protocols to use.

Recent years have seen this issue further exacerbated by the emergence of mobile devices and the IoT. We now have discovery protocols specifically tailored to work in scenarios that focus on low energy or proximity communication, such as Bluetooth, ZigBee and Wi-Fi Direct. Other protocols were adopted as general purpose solutions (e.g. Zeroconf [34]), or to address specific market segments (e.g. UPnP/DLNA [35]). Similarly, novel discovery protocols have also emerged to support recently proposed ICN architectures [14,36,37]. With the growing adoption of devices in these different areas, the fracture and lack of integration does not seem solvable through the emergence of a one size fits all discovery protocol.

At the core of this interoperability issue stands the fact that different protocols are subject to the utilization of different identifiers, resulting in multiple representations of devices and, as a consequence, different challenges to users. As a result, it becomes complex to realize discovery of devices and services in protocol-independent environments, even though such a solution would offer supportive measures in regards to interactions across different telecommunication technologies, and between users and the devices. Furthermore, the characteristics present in different IoT scenarios (i.e. the utilization of gateways and different protocols as illustrated in Figure 3.10) provide disparate challenges in regards to service and device discovery. For example, gateways are important in an environment where multiple network architectures are interconnected and interoperate, relying on supportive mechanisms such as the ones defined in Section 3.4.

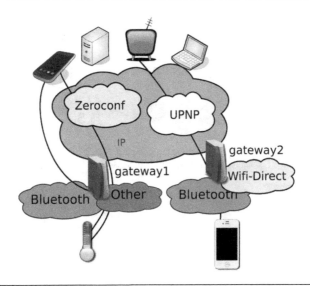

Figure 3.10 Discovery scenario involving multiple services, consumers and gateways.

The role of service discovery in any given network is usually twofold: to discover *nearby* devices and to enumerate services provided by each device. In this particular scenario, these roles are embodied by service consumers that need to find available services (either directly or through the gateways that advertise them), and services need to be advertised regardless of the intermediate steps involved (e.g. one or more gateways).

The challenges within this context derive from the existence of multiple protocols and multiple intermediate gateways. To concretely identify these challenges we iteratively split the previous scenario into four distinct cases.

Case 1: When considering the typical distributed IoT applications, a representative scenario starts as an array of sensors (Figure 3.11), from which we want to discover the available sensors using protocol *P*1 (e.g. Bluetooth or ZigBee). In this example a discovery protocol would fulfill the role of finding the services associated with those sensors, but it is also possible that the client only wants a specific sensor (S1), or a subset of all sensors (e.g. S1+S2 but not S3).

Case 2: To extend the previous scenario, and allow exposing the sensor(s) onto other protocols, a gateway would be added (Figure 3.12) that advertised the services over two new

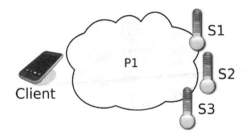

Figure 3.11   Case 1: Communicate with (sub)sets of sensors.

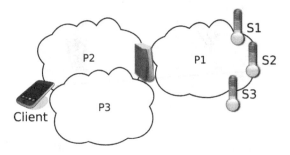

Figure 3.12   Case 2: Extending discovery to other protocols using gateways.

protocols *P2* and *P3*. This opens up possibilities for clients that support multiple protocols to take advantage of their simultaneous utilization, switching protocols as they perceive doing so to be beneficial. However, this can only be achieved if they hold the capability to identify both services being advertised as being the same service.

Case 3: By introducing additional gateways to widen the coverage range for one of the protocols (i.e. mobile sensor scenarios, or simply for redundancy purposes) (Figure 3.13), we compound on the challenges presented in case 2. The problem of introducing multiple gateways that map onto the same protocol (*P3*) is the risk of mapping the same service/sensor multiple times, potentially confusing the clients using that protocol.

A potential solution for this issue would require gateways communicating among themselves to exchange service rosters, to avoid duplication of service advertisements. However, this kind of approach creates a strong emphasis on the knowledge detained by each gateway, and on gateway coordination, both

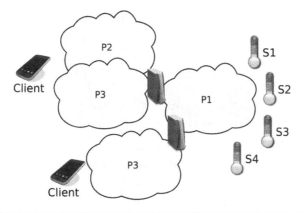

Figure 3.13   Case 3: Gateway replication.

of which are outside our defined scenario constraints: we assume that for these scenarios there is no centralized gateway management and any two gateways might not be able to communicate directly.

Case 4: One final case (Figure 3.14), derived from case 2, is when the client supports multiple protocols (and technologies), one of those being the protocol used by the sensors, in which case the client may bypass the gateway entirely, and be presented with a service it cannot identify as being the same. While it may seem the gateway is unnecessary (for that particular client), *P2/P3* may provide wider coverage than *P1*. More importantly, the client should be able to realize that the services being advertised over protocols *P2* and *P3* (i.e. advertised by the gateway), are the same services being advertised over *P1*.

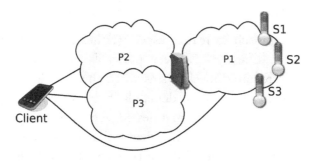

Figure 3.14   Case 4: Discovery consistency across gateways.

From the aforementioned cases, the following challenges are summarized:

- *Challenge* 1: The same service can be mapped to one or more devices; this may be a subset of all available devices
- *Challenge* 2: Service clients need to relate services advertised over different protocols as being the same service.
- *Challenge* 3: The replication of gateways should not break the previous point, even if the gateways are not aware of each other
- *Challenge* 4: Service identification is consistent between the originally advertised service and the mapped services advertised by the gateways.

3.5.2 Interoperability Approaches

In order to address this fractured interoperability environment, a mechanism to aggregate discovery procedures across different technologies becomes necessary. The cornerstone of such a mechanism has to rely on a universal identification system, or namespace, allowing the different entities to be recognized despite the utilization of disparate protocols. Approaches employing this mechanism need to support the following aspects:

1. Applications and services need to be able to identify the same devices from the results provided by different discovery mechanisms and protocols.
2. The approach needs to be able to leverage different technologies, despite the underlying technology associated with a device that was previously discovered and recognized.

A common approach to realize aspect (1) lies in the simultaneous deployment of different discovery protocols. This allows entities to employ the required mechanisms when the situation (i.e. protocol or technology being used) demands it. However, this approach allows for discrepancies to be manifested, such as wrongfully identifying the same device differently when different discovery protocols are used, or when the detecting entity is able to reach the detected device via different links. In this way, it becomes necessary to provide identity aggregation measures (i.e. a protocol agnostic identification mechanism).

In regards to aspect (2), the ability to employ different technologies is becoming a natural aspect in many different scenarios, with the proliferation of multi-interfaced devices (i.e. smartphones). This conveys the ability to deploy optimization measures, where the best connection is selected for receiving and/or sending a particular part of information belonging to an ongoing or new data flow. Combining this with aspect (1), new scenarios become available where reaching out to a specific device can be done using an alternative communication access. This can be the result of optimization procedures leveraged by user mobility or energy efficiency mechanisms. However, in order to reach a previously discovered device via a new link, it becomes necessary to identify the device as being the same, despite two different accesses (and potentially, different protocols, architectures or discovery protocols) being used. This emphasizes the need for a protocol agnostic identification layer, which is able to operate independently of protocol or technology, and is perceived as a unified discovery namespace.

To achieve this, a number of alternatives are available. The first is to assume that all gateways are connected, and can orchestrate the discovery results among them. However, this requires communication, not to mention compatibility, between gateways. A second approach is to promote a protocol,[10] or set of compatible protocols, across all architectures. For example, Apple uses the Zeroconf protocol over Bluetooth,[11] enabling easier mapping of this protocol in networks that combine IP networks and Bluetooth, but it is not compatible with legacy Bluetooth discovery protocols. Finally a third option, which will be the focus of our work, as demonstrated by [38] (but not applied there to the ICN use cases), is to reuse existing protocols while keeping a common identification namespace across all of them, relaxing the need for gateway synchronization.

---

[10] https://google.github.io/physical-web/
[11] https://developer.apple.com/bonjour/

3.5.3 Cross Architecture Resource Discovery

Discovery procedures across different protocols, technologies and architectures can be achieved through the provisioning of a common framework for device identification, which can be referenced in each protocol, technology or architecture.

Rui Ferreira et al. [38] hold two properties that are particularly relevant for interconnecting IoT with ICN: it retains compatibility with existing discovery protocols (Zeroconf, Bluetooth, UPNP), and enables the use of a common identifier namespace across these protocols. This is achieved through various encodings of a Universally Unique Identifier (UUID) namespace across these protocols. In this framework, each entity holds an asymmetric key pair, which has a UUID associated with it, generated from the public key using a hash function. The first component ensures that the identifier ownership belongs to the holders of the key pair, whereas the UUID provides a unique device identifier denoted as the *EntityID* (EID) that can be used by different protocols, technologies and architectures by converting the hash into a usable format.

The EID namespace is meant to map generic identifiers, the EIDs, into protocol-specific Entity Locators (ELOC). The purpose of the EID is to uniquely identify entities in the network, using a single identifier that can be used in various existing discovery protocols. Any EID is a version 5 UUID generated from an SHA1 content hash, according to RFC4122 [39],[12] generated from a piece of content:

$$EID = UUID(SHA1(content)) \tag{3.1}$$

The ELOC is a locator associated to the transport protocol over which a discovery protocol operates. Examples of ELOCs are MAC addresses (in Bluetooth SDP), IP addresses and ICN names. The ELOC is usually represented as a tuple:

$$ELOC = ProtocolFamily/ProtocolLocator \tag{3.2}$$

The mapping between EID and ELOC is a **M:N** relation, meaning that an EID can be associated with multiple network locators, and conversely any device in the network can advertise multiple EIDs.

---

[12] Using the fixed namespace identifier 166266d3-a4b9-4886-9cb3-6d53d3928d68

Since EIDs are generated from content hashes, one of the capabilities provided by EIDs is to tie a public key to a specific EID (similar to HIP [40]). This generates EIDs whose ownership can be verified based on public key cryptography challenges, e.g. using SSL or another appropriate protocol. For a public/private key pair (K^1, K^{-1}), the EID is generated using the public K^1. And all advertisers of this EID should hold the corresponding private key K^{-1}.

Since any EID can be mapped to multiple ELOCs and vice-versa, the holder of the private key can be a single device in the network, or multiple devices that share a common private key.

In addition to defining the namespace, [38] also provides an implementation for discovery functions that work in multiple protocols (Bluetooth SDP, Zeroconf, UPnP). The provided functions abstract only the naming of devices, as defined by the underlying discovery protocol. Applications still have to call protocol-specific APIs to parse device locators or for connection establishment. Its purpose was never to create an abstract discovery API, but to offer additional functionality through the creation of the namespace. This is a requirement for interoperable network architectures, which beyond abstracting the protocol, actually need to abstract the underlying network architecture.

3.5.3.1 Discovery Models: Extending EIDs as Service Identifiers The EID framework already provided us with the fundamental principle that we considered necessary to support this kind of IoT scenarios—the ability to identify the same service as being distributed or multihomed—however it lacked two main capabilities to make it immediately useful in IoT scenarios:

1. It did not provide service details, which made it hard to discover services without assuming static configuration (e.g. static service port) or incur additional discovery steps.
2. It worked strictly at the service client and advertiser, meaning that service advertisement would not transpose gateways, a fundamental concern for IoT scenarios.

In this section we describe our contributions towards the described scenarios, how to generate EIDs for distinct cases and how to implement gateways that take advantage of this. Afterwards in Section 3.5.3.7 we discuss how to put each of these capabilities in practice.

3.5.3.2 ELOCs as Service Descriptors Discovering services instead of devices implies adding the capability to map additional information to an EID, particularly service location information that the client can use to connect to the service. As such, the first step to extend the use of EIDs as service identifiers in the network involves updating the definition of the ELOC to include additional information. The ELOC is thus extended from its original form (Section 3.5.3, Eq. [3.2]) to a three element tuple with an additional (optional) element designated *ProtocolChannel*.

$$ELOC = ProtocolFamily/ProtocolLocator$$

$$[/ProtocolChannel] \qquad (3.3)$$

The *ProtocolChannel* holds the service information as returned by the underlying discovery protocol. This element is optional to retain compatibility with the case where the EID does not map into a service. For Bluetooth the *ProtocolChannel* holds the Bluetooth channel, for IP-based discovery the TCP or UDP port, and for ICN it can hold a segment of the content name. At the moment we do not include any additional service metadata in the ELOC, because it would risk breaking support for other protocols.[13]

3.5.3.3 Generating Non-authoritative EIDs The use of EIDs as service discovery identifiers opens the door for further extensions to the EID generation process itself. Originally, an EID was built using a content hash of a public key (Section 3.5.3), for the purposes of authentication. However, it makes sense to consider generating it from alternative sources of content. Our focus in this particular work is on generating EIDs from service-type identifiers. The advantages of doing this are that it becomes possible to use the same service-type identifiers, regardless of the discovery protocol being used. In doing this, the capability to authoritatively authenticate the owner of an EID is intentionally sacrificed, since any device can advertise any service type, by design.

The same method could be used to generate an EID from other types of authoritative names (e.g. DNS names or ICN names), but

---

[13] This is the minimal amount of information to instantiate a service description, but others could be included as well (e.g. service description)

clients and services would have to agree on an external method to verify ownership of those EIDs.

To generate an EID from a service type we apply the EID generation function as stated previously, using a service type as the content argument, e.g. for a printer service *EID('printer')* is *3fcfc044-ceb4-52d9-a6e8-90da061fb9c2*.

The source content for the EID is up to the service advertisement. For example, the service type of the *original* discovery protocol used by sensors could be leveraged here, across protocols.

As this hashing operation is non-reversible, (i.e. it is not possible to obtain the human readable service type from the hash), a similar requirement for other discovery protocols that use UUIDs is to keep a mapping table of well-known service types into matching identifiers (e.g. UUIDs or ports). This kind of approach is already used in ICN protocols where identifiers are the result of hash functions (e.g. PURSUIT). As any device can advertise multiple EIDs, consequently it can advertise as many services as necessary, as is the case for most service discovery protocols.

As pointed out, EIDs generated in this manner cannot be authenticated. However, any device can advertise multiple EIDs, and as such it can advertise an EID that corresponds to a public key along with EIDs for its services.

If a client needs to find a specific service type but only for the holder of a specific public key, it must look for locators that simultaneously map to the $EID(K^1)$ and the EID(SrvType).

Another method to generate the EID for this purpose that warrants consideration is to combine both pieces of data into a single EID. An example of such mapping could be

$$EID^* = UUID(SHA1('srvtype') + SHA1(K^1)) \qquad (3.4)$$

However, we refrain from using this type of approach as it prevents devices from discovering EIDs in the network using only the key, or only the service type. The advantages of such an approach also seem negligible, since at most we would avoid publishing one additional service to the network.

3.5.3.4 EID-Aware Gateways   One important contribution of defining EID-aware gateways is that the EID representation is a valid

UUIDv5 on all supported protocols, meaning that gateways that are aware of this can simply reuse the identifier onto the mapped services.

It is important to realize that for protocols that already use UUIDs for service types (e.g. Bluetooth), there should be no collision with existing services, because UUIDv5 keeps 4 different bits for the version that will differ from other UUID generation versions. Even if there are other services in the network using version 5 UUIDs as service-type identifiers, the probability of collision should still be low, even if they hash the same content, since part of the UUID is generated from the namespace identifier [39].

As such, an EID-aware gateway could clearly distinguish between the services in the network that can refer to an EID and those that cannot. If a gateway found a service being advertised with an EID, it could create a new service in another protocol reusing the same identifier, while other services would be ignored (not mapped) or mapped using other solutions (e.g. [41]), without the benefits of an EID.

3.5.3.5 Implementation To enact these changes we extended the original implementation from [38]. The base design concept provided the necessary mechanisms for mapping the namespace of EIDs to multiple protocol-specific namespaces (the ELOCs). This base design consisted of a set of operations, comprising an API that provides support to existing discovery protocols. In the extensions presented in this section, those APIs were enhanced, while remaining backwards compatible, as were the wire encodings of the EIDs over the different protocol messages. We now enumerate the framework functions and the necessary changes.

1. **Publish-Entity(EID, [ELOC])**: In its original form, this procedure creates the protocol-specific identifiers and advertises them by publishing EID information to the network, and allowing other devices to discover them. It now accepts an optional ELOC parameter used to publish services as well.

2. **DiscoverEntities()** → [EID, protocol, ELOC]: This function is used to discover nearby entities and return the associated locators. This was extended to include service description information, as part of the ELOC.

3. **Entity-to-Locator(EID)** → [ELOC]: This function returns the locators for the corresponding EID. This is used to find entities

with a specific EID. As in the previous case, this was modified to include service description information as part of the ELOC.

4. **Locator-to-Entity(ELOC)** → [EID]: Translates a locator from a specific discovery protocol into a EID. This function remains unchanged.

In addition we implement a prototype of a service gateway that maps EIDs across protocols. The gateway retains compatibility with the previous process of injecting the EIDs as ancillary discovery information, meaning it is backwards compatible with [38]. For the considered ICN architectures, the EID can be applied as part of existing identifiers. In NDN it can be used as a name segment, and in PURSUIT it can be converted back into a partial hash (i.e. the Scope Id).

3.5.3.6 Cross-Protocol Support in Existing Technologies It is important to acknowledge that there is wide a variety of protocols that serve the purpose of service discovery [42], depending on the underlying network protocols, applications or specific services being used, e.g. mobile devices use Bluetooth SDP or more recently Wifi-Direct. Multimedia devices and services use UPnP/DLNA [35], other applications employ general purpose protocols like Zeroconf [34] and sensor networks employ discovery functions embedded within the network protocol itself (e.g. ZigBee).

Solutions that enable crossing the technology link boundary usually appear in the form of convergence gateways that map service discovery between two or more protocols, such as [41,43,44]. Bridging technologies implies discovering services using one protocol and publishing the corresponding entries in another, a practice which is common in IoT scenarios where distinct sensors are exposed and aggregated through intermediate gateways.

The advantage of these solutions is that the mapping process is transparent to service consumers and service advertisers, where the gateway is usually part of the network infrastructure. To carry out the mapping, the gateway must hold knowledge of the types of services that it will map, with no provision to map unknown service types or protocols. In the process, gateways may obfuscate the real identity of the service, making it hard for client applications to take advantage of multihoming capabilities.

A complementary approach, that we assume here [38], is that our extended framework allows the retention of a common identification scheme (EID) across discovery protocols and also allows the realization of the proposed scenarios (Section 3.5.1), and might even reduce the complexity of the gateways for IoT scenarios in general.

3.5.3.7 Addressing Identified Challenges   The following sections address how our extended framework maps with the individual cases and challenges identified in Section 3.5.1.

3.5.3.7.1 Challenge 1: Services that Spawn Multiple Devices   The two primary types of identifiers in service discovery are device identifiers and service types, one of which is used to identify devices to express their capabilities. However, EIDs can be mapped to multiple devices simultaneously, and a device can advertise multiple EIDs.

Following from the scenario in Figure 3.12 and applying it to a SmartHome situation [45], it can be considered that all sensors advertise the same EID that identifies them as being part of single household (*EID_H*), and *S1/S2* advertise an additional EID(*EID_L*) that identifies them as being in a specific part of the house.

This ability to target a subset of services, even if they are of the same type, is not very common in generic discovery protocols, with the exception of application-specific metadata or protocols where service discovery is implemented as part of message routing (e.g. ZigBee or other custom sensor routing protocols). The concept is similar to that of anycast in IP, except that here it is handled as part of discovery functions, rather than routing.

This approach implies that if security is necessary and EIDs are generated from public keys, then each sensor would need to hold the private keys corresponding to the advertised EIDs. In addition, the application advertising the multiple EIDs must use the correct key when challenged for authentication, which may implicate deploying the same service across multiple ports. If no security is necessary then sensors may advertise using non-authoritative EIDs (Section 3.5.3.3).

3.5.3.7.2 Challenge 2: Cross-Protocol Discovery The general goal behind the EID framework is to provide an abstract discovery namespace that can be used over multiple protocols. As presented in [38], it defines a common namespace, based on UUIDs that could be used to identify devices, and is extended here to identify services. The advantage of this solution is that the EID can be used consistently across protocols, meaning it is possible to identify the same service as consumers move across environments.

This allows a discovery client to switch network protocols (e.g. from Wifi to Bluetooth) according to power saving or bandwidth policies for the application (Figure 3.15). The EID works as a way to redis-cover known services.

3.5.3.7.3 Challenge 3: Retaining EID Information across Gateways As illustrated in Figure 3.13, adding multiple gateways that map the same service into the same protocol may cause services to appear mul-tiple times to a client. To retain EID information across gateways, we outlined in Section 3.5.3.4 how an EID-aware gateway should behave so that clients can still identify multiple service advertisements as per-taining to the same EID.

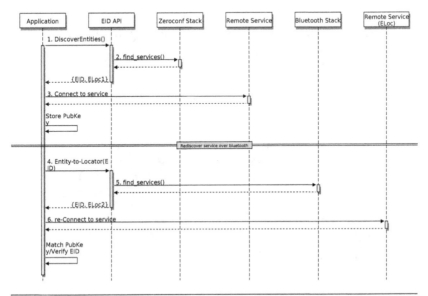

Figure 3.15   Client rediscovers a known service over Bluetooth/Zeroconf using the EID.

Our concern was to avoid erroneously mapping services that use ambiguous service identifiers that can be mistaken for EIDs. However, these two types of protocol identifiers can be distinguished since the UUID service identifier generation for each is based on different methods.

This means the same gateway can map EID services as through this framework, while using other methods to map the remaining services. Naturally this is contingent on having support at the gateways, but we have effectively extended [38] to work along service mapping gateways.

3.5.3.7.4 Challenge 4: Service Identification Consistency The mechanism proposed in this chapter is mainly geared towards scenarios were there is no central management and gateways work independently from each other. In real-life scenarios small sensors tend to offload work to proprietary gateways, due to processing power constraints or in order to minimize energy consumption. It is not unexpected to face situations where sensors do not hold the capabilities to implement asymmetric cryptography for security, cannot be modified to enforce the use of an EID or operate using closed protocols.

However, the cases we identified are equally applicable to chained gateways (Figure 3.16), if we want to hold consistency across multiple gateways (e.g. if *P1* is a closed protocol). This aspect becomes more relevant if we consider the increased number of gateways for extensibility and backwards compatibility.

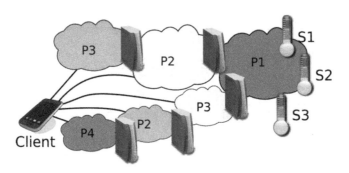

Figure 3.16 Discovery consistency across a chain of gateways.

3.6 Information-Centric Exchange Mechanisms for IoT Interoperable Deployment

To mitigate the impact of information "silos" limited to specific IoT solutions that are not able to interact with other IoT solutions, we bring the interoperability mechanisms defined in Section 3.4 into an IoT environment enriched with the enablements provided in Section 3.3. IoT scenarios are generally characterized by their heterogeneity in terms of devices, protocols and, with growing potential, even architectures. As such, we designed an IoT scenario involving different networking solutions, namely the current IP architecture along with NDN and PURSUIT ICN architectures. Using the interoperability mechanisms, consumers can access information from sensors on each network architecture independently of its own architecture. The resulting framework was developed in a flexible way, allowing its integration with other supportive mechanisms, such as added security, mobility support or even novel discovery procedures, such as the ones proposed in Section 3.5.

3.6.1 Solution Overview

In Figure 3.17, the envisioned IoT deployment scenario, where sensors and consumers are deployed across different network architectures, is presented. Because of the widespread use of the IP architecture in the current Internet, it was chosen as the connecting point between two distinct ICN architectures (more specifically, NDN and PURSUIT). Nonetheless, a dedicated connection between the NDN and PURSUIT could also be added in our scenario to avoid messages being sent over the IP network. In the IP architecture, two different resource-constrained protocols for IoT were taken into

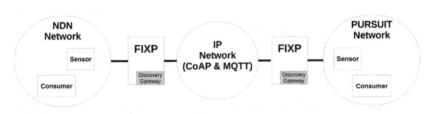

Figure 3.17 Information-centric exchange mechanisms for IoT interoperable deployment scenario.

consideration: (i) Constrained Application Protocol (CoAP) [46], due to its request/response communication model similar to NDN architecture, and (ii) MQTT [27] due to its publish/subscribe communication model similar to PURSUIT architecture.

To support FIXPs in discovering the sensors on each network architecture and mapping them across the remaining architectures, the discovery mechanisms proposed in Section 3.5 can be applied, by integrating them into the FIXP itself. As such, FIXP instances hold the same properties described in Section 3.5.1, for discovery gateways. Thus, sensors are advertised in other network architectures besides the one they are deployed in, with consumers seeing them as belonging to the same network architecture and accessible via supported protocols.

After discovering existing sensors, consumers can acquire their sensing information using two different approaches: (i) via polling mechanisms, where information is continuously requested by consumers or (ii) via publish/subscribe mechanisms, where information is subscribed once and pushed whenever an update occurs.

Regarding the presented scenario, polling and publish/subscribe are respectively achieved:

- **In the IP network architecture,** by using CoAP and MQTT protocols;
- **In the NDN network architecture,** by using the native NDN mechanisms and the NPSN protocol as defined in Section 3.3.2;
- **In the PURSUIT network architecture,** by subscribing, fetching and unsubscribing the information each time, and by using the native PURSUIT mechanisms.

Finally, since FIXPs are seen as the "real" information provider on each network architecture, messages destined to other architectures are sent towards a FIXP which translates the messages to be compatible with the destination network architecture, replicating the desired communication model across each architecture.

3.6.2 Scalability

The FIXP itself holds no state outside of the mappings it configures and the ongoing requests and responses. This means both horizontal and vertical scalability strategies are possible. Vertical scalability is achieved through the addition of computing resources and bandwidth to the FIXP, allowing it to handle an increasing number of connections.

Horizontal scalability requires the deployment of new FIXP instances to handle other connections in parallel. Usually this requires synchronization mechanisms to manage state across multiple instances. Alternatively, an agreed-upon partitioning strategy across all instances to divide resource handling responsibilities could be used. In structured networks this is sometimes achieved through a load balancer that splits the load across multiple workers.

The introduction of multiple FIXPs, from horizontal scalability, or as the network grows to support more architectures, may result in the introduction of non-optimal paths. When discovery solutions use broadcast or multicast, this may cause flooding throughout the network. Given the common unique identifier naming scheme seen in Section 3.5 it becomes possible for two FIXPs to detect they are mapping the same resource and follow contention strategies to avoid duplication of mappings. This may not completely avoid the broadcasting of the same information over network loops, but it provides a mechanism to minimize its impact upstream.

If discovery information flooding is to be avoided, centralized strategies for managing resource mappings at the FIXP can be considered, i.e. to delegate the creation of the mappings in the FIXP to a centralized entity, which avoids the duplication of mappings in the network.

3.6.3 Example Smart City Scenario

We consider an example scenario involving a smart city where sensors are used to monitor and manage its facilities. On an initial development stage, the deployed IoT devices in the city were IP based supporting the CoAP and MQTT protocols. The acquisition of new sensors for the watering system and for weather forecasting, which, respectively, only supported PURSUIT and NDN network architectures, led to a heterogeneity of deployed network architectures, similar to the one presented in Figure 3.17. However, intelligent agents (i.e. the consumers) on each architecture were unable to directly obtain the information from sensors outside their own architecture, since each architecture operated based on different design and communication models. Thus, to allow intelligent agents to access the information from the sensors deployed in other architectures, the contents need to be requested in a manner compatible with the mechanisms supported by their own architecture, using an interoperability entity responsible for the conversion of the messages to others compatible with the mechanisms supported by the sensors' network architecture.

With the solution presented in this section, the intelligent agents deployed in one of the ICN architectures can fetch the sensing information of a sensor on the other ICN architecture, via polling and publish/subscribe communication models depending on their capabilities or preferred way of obtaining the information. The signaling when fetching information via polling mechanisms from the PURSUIT and NDN architectures are, respectively, presented in Figures 3.18 and 3.19, and via publish/subscribe mechanisms are presented in Figure 3.20 when the request is originated in the PURSUIT architecture and in Figure 3.21 when the request originates from the NDN architecture.

In Figure 3.18, the PURSUIT consumer, desiring to access the sensor information, starts by subscribing its interest with the *PURSUIT Rendezvous* (message 1), which notifies the FIXP to start publishing the desired information (message 2). The FIXP translates the received message into a *CoAP CON* message, carrying the request for the desired information, to be sent towards the second FIXP (message 3). In turn, by receiving the *CoAP CON* message, the second FIXP obtains the desired information from the sensor itself (or

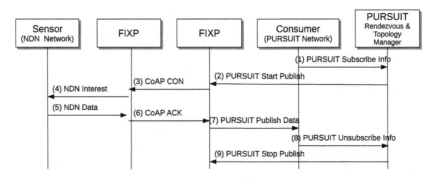

Figure 3.18 *PURSUIT-CoAP-NDN*: Consumer in PURSUIT requesting information from sensor in NDN via polling mechanisms.

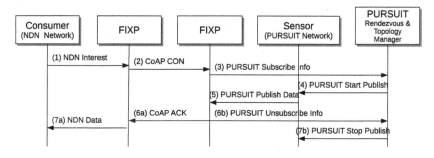

Figure 3.19 *NDN-CoAP-PURSUIT:* Consumer in NDN requesting information from sensor in PURSUIT via polling mechanisms.

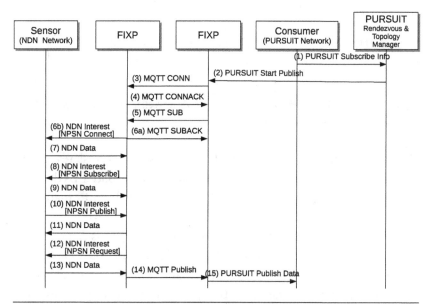

Figure 3.20 *PURSUIT-MQTT-NPSN:* Consumer in PURSUIT requesting information from sensor in NDN via publish/subscribe mechanisms.

any entity containing a cached copy) via NDN protocol (messages 4 and 5), sending it to the first FIXP by piggybacking it in *CoAP ACK* message (message 6). The information is then sent to the PURSUIT consumer (message 7). Finally, since the sensor operates based on polling mechanisms, the consumer unsubscribes its interest for the sensor's information (messages 8 and 9).

The opposite case is presented in Figure 3.19, with the NDN consumer starting by issuing an *NDN Interest* message expressing its interest in the sensor's information (message 1). Upon reception

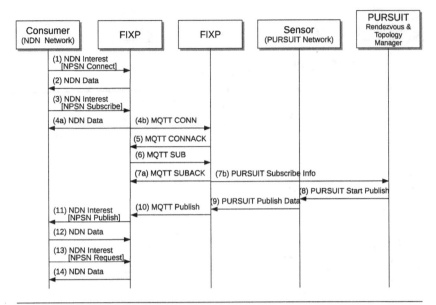

Figure 3.21   *NPSN-MQTT-PURSUIT.* Consumer in NDN requesting information from sensor in PURSUIT via publish/subscribe mechanisms.

of this message, the FIXP, as in the previous case, translates it into a *CoAP CON* message to be sent towards the second FIXP (message 2), triggering the second FIXP to subscribe the information in the PURSUIT network architecture (messages 3 and 4). After being provided with the desired information (message 5), it is piggybacked in a *CoAP ACK* message (message 6a) and then in an *NDN Data* message (message 7a), allowing the information to be delivered to the original requester. At the same time, knowing that the request was made via polling mechanisms, the second FIXP unsubscribes its interest for the sensor's information (messages 6b and 7b).

In Figure 3.20, the PURSUIT consumer subscribes the desired information with the *PURSUIT Rendezvous* (message 1), which notifies the FIXP to start publishing the requested information (message 2). In this case, since a publish/subscribe behavior is preferred, the received message triggers the FIXP to establish an MQTT connection with the second FIXP (messages 3 and 4) and afterwards the message is translated into a *MQTT Subscribe* message (message 5). Upon reception of the subscribe message, the second FIXP obtains the desired information from the NDN sensor using the NPSN protocol as defined in Section 3.3.2 (messages 6b–13). The information is encapsulated in an *MQTT Publish* message (message 14) to be delivered to the first FIXP, which in turn encapsulates the information in a *PURSUIT Publish Data* in order to be delivered to the original requester (message 15). When an update to the sensing information occurs, the new information is pushed to the PURSUIT subscriber following the same procedure defined in messages 10–15.

In the inverse case, as depicted in Figure 3.21, the consumer in the NDN architecture starts by establishing an NPSN connection with the FIXP (messages 1 and 2), after which it sends an *NPSN Subscribe* message (message 3) to subscribe the desired information. This message triggers the FIXP to establish a MQTT connection with the second FIXP (messages 4b and 5), after which it translates the received subscribe message into a *MQTT Subscribe* message (message 6). The second FIXP, receiving this message, subscribes the desired information with the *PURSUIT Rendezvous* (message 7b). The *PURSUIT Rendezvous* notifies the PURSUIT sensor to start publishing the information (message 8). The information is then sent towards the FIXP (message 9), which encapsulates the information into a *MQTT Publish* message (message 10) to be delivered to the first FIXP. Finally, the FIXP notifies the original requesters that it has data to be pushed (message 11), triggering the latter to explicitly request the desired data (messages 13 and 14). Subsequent updates to the sensor's data are pushed to the consumer in the NDN architecture following the same procedure as defined in messages 9–14.

3.6.4 Evaluation

The presented solution was evaluated in the scenario depicted in the previous section. For that, a proof-of-concept prototype integrating the information-centric exchange mechanisms was implemented, supporting the previously mentioned architectures (for supporting NDN and PURSUIT network stacks, NFD[14] and Blackadder[15] were used, respectively).

---

[14] Named Data Networking Forwarding Daemon (NFD): http://named-data.net/doc/NFD

[15] Blackadder (FP7 PURSUIT): http://www.fp7-pursuit.eu/PursuitWeb/?page id=338

For this evaluation, we calculated the average rate and size of messages exchanged by sensors based on captured data from the Smart Cloud of Things project (SCoT).[16] Thus, the emulated sensors sensed the environment with a 100 ms periodicity, generating sensing data with 143 byte size.

This scenario was deployed in an OpenStack Platform, with each entity running on a virtual machine with two 3.33 GHz CPU cores and 2 GB of RAM.

3.6.4.1 Transit Network Impact This evaluation aimed to assess the impact on the transit network whenever a consumer on one of the ICN architectures fetches the sensing information of a sensor on the other ICN architecture, as in the signaling depicted in the previous section. Results for the fetching times are presented in Figure 3.22 (with reference fetching time values for each architecture, in our evaluation scenario, being presented in Figure 3.23) and the overhead introduced on each architecture is shown in Table 3.2.

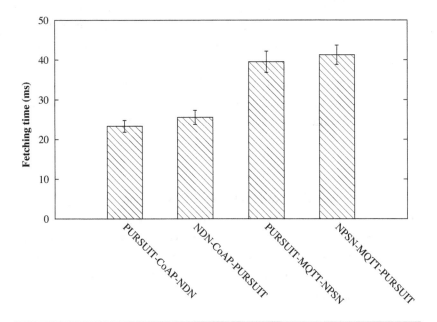

Figure 3.22 Fetching time for a single request.

---

[16] SCoT: https://atnog.av.it.pt/content/smart-cloud-things

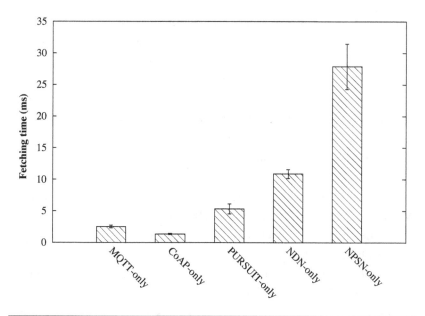

Figure 3.23   Fetching time for a single request (reference values).

Table 3.2   Signaling Overhead on Each Network Architecture (For a Single Link)

	EXCHANGED INFORMATION (BYTES)	EXCHANGED MESSAGES
PURSUIT-CoAP-NDN	PURSUIT: 713	PURSUIT: 5
	CoAP: 264	CoAP: 2
	NDN: 667	NDN: 2
NDN-CoAP-PURSUIT	NDN: 686	NDN: 2
	CoAP: 271	CoAP: 2
	PURSUIT: 713	PURSUIT: 5
PURSUIT-MQTT-NPSN	PURSUIT: 490	PURSUIT: 3
	MQTT (+TCP): 567 (+478)	MQTT (+TCP): 5 (+7)
	NPSN: 2154	NPSN: 8
NPSN-MQTT-PURSUIT	NPSN: 2428	NPSN: 8
	MQTT (+TCP): 599 (+478)	MQTT (+TCP): 5 (+7)
	PURSUIT: 490	PURSUIT: 3

From the results we observe that the fetching times, when interoperability is required, are higher compared to the reference values of fetching the sensing information when both sensors and consumers are on the same architecture. In fact, since the FIXP needs to obtain the sensing values on behalf of the original requester on the different supported architectures, when interoperability is required, the fetching time of individual requests for each network architecture will contribute to the total fetching time.

In terms of overhead, in the polling cases, transporting the information via CoAP demonstrated it to be the protocol that required fewer messages (i.e. 2 messages) and exchanged bytes (i.e. 264 and 271 bytes for the *PURSUIT-CoAP-NDN* and *NDN-CoAP-PURSUIT*, respectively). Similarly, the NDN protocol only required 2 messages to fetch the information, but since more metadata goes into the NDN packets, the number of exchanged bytes increased to 667 bytes for the *PURSUIT-CoAP-NDN* case and 686 bytes for *NDN-CoAP-PURSUIT*. In the PURSUIT network, for both cases, the higher number of exchanged messages and bytes (i.e. 5 messages and 713 bytes) was registered.

In the publish/subscribe cases, transporting the information via PURSUIT required fewer messages and exchanged bytes (i.e. 3 messages and 490 bytes). In contrast, MQTT was the protocol that required more messages (i.e. 12 messages), a higher number due to the need for TCP control signaling, and NPSN was the protocol that required more exchanged bytes (i.e. 2154 and 2428 bytes for *PURSUIT-CoAP-NPSN* and *NPSN-CoAP-PURSUIT*).

3.6.4.2 Major Lessons   The flexibility of the mechanisms presented in Section 3.4 allowed their applicability on an ICN-enhanced IoT use case without requiring modifications to its core.

In addition, the proposed solution enabled two different network architectures to be interconnected by a third architecture. However, each architecture hop had an impact on the performance, increasing the time required to fetch the desired information.

Also, compared with the web browsing scenario presented in Section 3.4.3, since the information retrieved by the sensors is usually in its raw state, it does not contain identifiers to external resources (i.e. URIs for other resources) and therefore, analyzing the content by the FIXP to replace existing URIs was not applicable, which allowed a decrease in message processing time in the FIXP process.

Notwithstanding the previous points, as stated in Section 3.4.4.2, the middleware solution introduced by the FIXP has a set of issues that are also applied to an IoT interoperable deployment. These issues are currently being tackled in order to mitigate their impact in the communication between endpoints in different network architectures. Next, we briefly describe the key issues.

- **Anchoring-related issues:** since message conversions is performed at the FIXP, this entity is seen as an anchor point for communications that cross different network architectures. Thus, non-optimal routing or overloading of the FIXP entity is likely to occur. While the first may impact the delay in the communication between the consumer and the sensor/broker, the second may impact the scalability of the solution. In addition, a response to a previously converted request message must follow the reverse path of FIXPs to be delivered (e.g. the FIXP that handled the Interest message in NDN must be the FIXP that handles the Data message; otherwise, the delivery of the response message may fail since no entry in the PIT existed). In the IoT scenario presented in this section, sensing information was not critical and, therefore, a delay in the delivery of the information was acceptable. However, in more critical scenarios where information is time sensitive, this may become an issue and, therefore, further optimizations in the proposed solution may be required.

- **Receive complete messages:** before the FIXP is able to convert the received messages, it needs to receive the complete message. As such, fragmented messages will be queued in the FIXP before being converted towards the destination network architecture, which may introduce an additional delay in the communication between endpoints in different architectures. In our specific IoT scenario, this was not an issue since the IoT content was significantly small and it was received in a single packet. However, if we take into account other types of IoT applications (e.g. video surveillance) on which the size of content is larger, an increase in the time to deliver the message to the destination endpoint is likely to occur.

- **Creation of identifiers for resources:** whenever an identifier for a resource is composed of a hash (or other element) of the content itself (e.g. self-certified identifiers), it is not possible for the FIXP to strictly follow that assumption whenever it creates identifiers. This is because the creation of those identifiers is performed before the content itself is received and thus, the content is unknown to the FIXP. To mitigate this issue, the identifier could be updated after the first conversion, invalidating the previously advertised identifier. In our IoT scenario, identifiers pointed to mutable data and, therefore, identifiers were not self-certified.

- **Security considerations:** security and privacy aspects may not be implemented in the same way across the different network architectures. As such, during the message conversion the FIXP needs to replicate the same security level using the mechanisms provided by each network architecture. However, this may imply that the FIXP infrastructure is a trusted entity for all participants (e.g. network providers, resource providers and clients). For example, in our specific IoT scenario, the FIXP acts on behalf of sensors/brokers while delivering the sensing information to the consumers and, therefore, whenever required, secure communications must be replicated using relevant keys and certificates. As a result, resource providers may need to delegate trust to the FIXP infrastructure.

3.7 Management of Interoperable IoT Deployments

IP host-centric networks have had years of evolution and several network management tools and protocols have been developed. On the other hand, ICN architectures are still in their infancy. However, it is clear that for ICN solutions to be widely deployed, network management is a key challenge to be addressed along with the design and development of the core network functionalities [47]. The shift in terms of the focus of the networking functions introduced by ICN also requires management to be rethought. While elements such as content integrity, authenticity, load-balancing, flow control and resiliency are deeply addressed at the core of the ICN architectures, other elements are more challenging, such as:

- How to name entities in a way that enables management across architectures
- How to ensure functionalities such as AAA when senders and receivers are decoupled
- How to enforce client-specific quality of service

Parallel efforts, such as [15], address the latter two items transparently, while the main focus of this work is the first item.

3.7.1 Device Provisioning

From Section 3.5 it has already been established that new devices must adhere to a common naming scheme, which enables consistent identification across different protocols. This can be done over several existing discovery protocols, but requires a consistent enrolling of new sensors into the network across different protocols (PURSUIT, NDN).

IoT deployments are expected to involve a large number of highly heterogeneous nodes, and as such it is expected that IoT deployments have autonomous capabilities, thus ensuring minimal human-related management tasks.

Sensor information is made available through different protocols according to the network architecture. Consumers and producers need to seamlessly discover each other, across different environments.

Existing solutions can be used to fulfill this purpose, such as ndn-autoconfig (which is part of the NDN Platform), local discovery approaches [48] and semantic discovery approaches [31]. Moreover, solutions involving more than one discovery mechanism should also be explored for more complex and complete scenarios.

For this particular deployment NPSN can also be used to fulfill this role, providing a reserved topic (secured using NDN authentication mechanisms [21]) for management functionalities (e.g. /SYS/discovery could be a topic reserved for discovering available topics).

As devices reach the end of life and are removed from the network, their enrollment to NPSN expires and the sensor data cached by the network (NDN routers) eventually expires according to cache policies.

3.7.2 Deployment Strategies

As IoT environments may vary in size and scope, one must consider how different approaches match different scenarios. The described components form an array of tools that can be arranged according to necessity.

The FIXP operates as a gateway across multiple network architectures. Device availability is determined from the discovery protocols available in different architectures. The FIXP then publishes discovery information over other network architectures. Transitively, a chain of FIXP instances share information about all sensors in a network, and all instances know about all sensors. Loops are avoided through the unique identification scheme from Section 3.5. The FIXP automatically creates mappings according to discovery information, instantiates the relevant endpoints and manages these mappings as devices become available or unavailable in the network.

The main advantage of this approach is that it avoids the use of centralized deployments. However, for other environments, the need for different deployment strategies arises, particularly for larger networks where the use of broadcast/multicast discovery protocols is limited by network topology.

An alternative approach is to consider a centralized (or hierarchical) entity that gathers information from the FIXP instances with neighboring sensors, and configures mappings across all instances in the network. This relieves the FIXP from management tasks (mapping creation, resource advertisement), and avoids some of the discovery propagation delays.

3.8 Conclusions

"Future Internet" has been one of the most chameleonic expressions ever applied in the telecommunications field. In the past, it has been used to distinguish the IPv6 and IPv4 versions of the Internet Protocol, as well as to refer to evolutions from the 3rd and 4th generations of mobile networks and, some years ago, it evidenced the prospects for the evolution and overall achievements of the Internet Protocol itself. With the advent of the 5th generation of mobile networks, and with discussions about Beyond-5G already starting, only two aspects are certain: i) evolution is an on-going process and the needs of today are not guaranteed to transition into the needs of tomorrow and ii) even if for a brief time, different alternative (or even competitive) technological deployments can co-exist simultaneously.

A key contributor to this plethora of technological needs, deployments and evolutions has been the application of communication capabilities to physical devices, in the realization of what is called the IoT. In this way, the provisioning of Internet services has increasingly being subjected to widely different scenarios and applications, evidencing, for the first time, the prospects of true massive communication, with tens of billions of smart devices connected.[17] It is clear that different aspects of communication provision will be deployed in a customized way, tailored to the requirements of the involved devices, architectures and generated information. This is already evidenced today, with different Machine-to-Machine communication protocols (such as MQTT and CoAP, used in this chapter) or even the recent competition between different Low-Power Wide-Area Network (LPWAN) protocols, such as LoRa and SigFox. This complexity and abundance of solutions is further exacerbated when the integration of other supportive Internet-based (and its underlying infrastructure-based) technologies come into play, further pushing the envelope of available connectivity solutions. If we add to the mixture the occurring changes in utilization paradigms, with the explosion of mobile users consuming (and now generating) high-definition video while on the move, the Internet is starting to see serious challenges to its ability to operate as the thin waistline between the application and the access technology layers that compose the hourglass model figure [49].

In this respect, Future Internet proposals have been in demand, and have paved the way for the definition of effective proposals, such as Information-Centric Networking, which not only provide comprehensive new ways to realize networking procedures, but actually encompass key properties in their intrinsic design, which in the case of the Internet Protocol were only added later on (e.g. security, mobility). But rather than being the definite solution to the upcoming challenges of tomorrow's overall global communications requirements, it has been hindered by several factors. To name a few: (i) ICN has been fractured in different architectures that operate in a disparate way (profoundly more different than, e.g. IPv4 and IPv6); (ii) despite providing intrinsic support for key operational

---

[17] Next Generation Mobile Networks: NGMN 5G White Paper, https://www.ngmn.org/uploads/media/NGMN_5G_White_Paper_V1_0.pdf

aspects such as security and caching, its application in important scenarios (such as the IoT) has never risen as one of the flagship deployment scenario cases; (iii) there are no plans for integration roadmaps, or even any kind of OPEX/CAPEX expectation, except for studies on Content Centric Networking cache performance [50] that provide a best guess for vendors; and iv) the current race towards 5G is divided between the emphasis on access network enhancements and the improvement and wide generalization of more flexible mechanisms for service provisioning.

In this way, this chapter evidenced three main aspects pertaining to ICN deployment, as a result of the challenging and continuously evolving communications environment of today. First, the application of ICN principles to IoT, albeit having received a fair share of research, has been portrayed as a secondary use case in regards to more obvious constructs that benefit from name-based and inherent caching mechanisms, such as content delivery and secure-based information dissemination. In this regard, we have showcased important opportunities and challenges to be addressed by ICN for IoT support in Section 3.2, complemented in Section 3.3 by a summary of the current efforts towards efficient IoT operation on top of ICN. Section 3.3 also presents and evaluates a Publish/Subscribe protocol on top of the Named Data Networking framework for an efficient support of IoT scenarios characterized by unpredictable event occurrences. Secondly, we explored the application and definition of inter-exchange mechanisms in Section 3.4, allowing different networking architectures (including ICN-based ones) to interoperate, evidencing translation performance and interfacing issues from different experimented architectures. Thirdly, Section 3.5 explored mechanisms assisting with resource discovery in multi-protocol environments, addressing different identifying challenges.

The final key contribution of this chapter, in Section 3.6, was the evaluation of the combined effort of the IoT-enhanced ICN framework operating in a heterogeneous networking architecture environment, deploying the presented interoperability approach (with management aspects reflected in Section 3.7). Despite the flexible integrated design of the individual components (which allowed them to operate jointly with minimal to no changes), the impact of the translation mechanisms evidenced the need for intrinsic interoperability, interfacing and incremental deployment mechanisms in new protocols and architectures, which is commonly an overlooked aspect in their origin.

3.9 Future Research Directions

Interoperability has always been an aspect associated with ICN. As a Future Internet technology, its name-based mechanisms have allowed the development of whole new kinds of communication scenarios and even the improvement of existing ones, though there have also been challenges concerning the execution of regular mechanisms existing in today's technologies. Nonetheless, such evolutions were realised unconstrained in regards to how they would impact the overall new architecture deployment. As such, ICN's operation across multiple network paradigms [2] not only has to address its integration into existing IP-based environments, but also must address different instantiations of ICN.

It is thus envisioned that ICN interoperability will increasingly become a key research focus, pasticularly because the IP-based world is not standing still: 5G is right around the corner with the introduction of a myrriad of new network-based mechanisms, allowing the network to encompass new communication scenarios and verticals, in a flexible, dynamic and cost-effective way. Will ICN become a client of a 5G-based architecture, or will it keep pushing unprecedent new utilisation use cases in Beyond-5G research? Albeit a decisively complex field, with the integration of heterogeneous aspects, devices and communication technologies, this work focused on an IoT scenario. IoT has generated an important production vector for Telecommunication and Information Technologies, establishing a base of upcoming efforts such as the Industry4.0 and connected vehicles. However, with 5G, and with ICN as well, many other new areas and verticals will impose their strict requirements and expose different kinds of services, further demanding more generically deployable interoperability solutions.

We hope that this chapter will contribute to the understanding of trends in the development of, and the challenges facing, next-generation networks, and pave the way for further innovations in this fast evolving arena in order to deliver the Future Internet.

Acknowledgments

In the scope of R&D Unit 50008, this work was financed by the applicable financial framework (FCT/MEC through national funds and when applicable co-funded by FEDER – PT2020 partnership agreement) with ref. no. UID/EEA/50008/2013. This work was also supported by the FCT Grant SFRH/BD/96553/2013.

Glossary

Constrained Devices: Devices that have limited resources (e.g. computational, communications, energy, memory), which are quite common in IoT environments (e.g. sensors, actuators).

Content Store (CS): Represents a table indicating the content cache in networking routers, in the NDN and CCNx ICN instantiations. Here, content is stored and its identifying name can be matched whenever requests reach that router.

EntityID (EID): Unique device identifier that can be used in various discovery protocols.

Entity Locator (ELOC): Locator associated with the transport protocol over which a discovery protocol operates.

Forward Information Base (FIB): Entity, in the NDN and CCNx ICN instantiations, that contains a set of entries that indicate through which interface certain content can potentially be reached.

Heterogeneous Networking: A telecommunications environment featuring the interconnection of different types of access technologies, protocols or networking architectures.

Information Freshness: Refers to the amount of time elapsed since a given piece of information was generated.

Multihoming: A device with the ability to be connected via multiple network interfaces.

Pending Interest Table (PIT): An entity, in the NDN and CCNx ICN instantiations, that keeps track of requests for specific content.

Rendezvous Point: A name resolution point, used in the PURSUIT ICN instantiation, which acts as a broker for the matching between the publication and subscription of content.

Topology Manager: Entity that monitors the network topology and signals changes to it, updating routing mechanisms in the PURSUIT ICN instantiation.

Acronyms

6LoWPAN	IPv6 over Low power Wireless Personal Area Networks
API	Application Program Interface
CAPEX	Capital Expenditure
CCN	Content-Centric Networking
CoAP	Constrained Application Protocol
CS	Content Store
DLNA	Digital Living Network Alliance
EID	EntityID
ELOC	Entity Locators
FIB	Forwarding Information Base
FIXP	Future Internet eXchange Point
HIP	Host Identity Protocol
HTML	Hypertext Markup Language
HTTP	Hypertext Transfer Protocol
ICN	Information-Centric Networking
ICNRG	Information-Centric Networking Research Group
ICT	Information and Communication Technologies
IoT	Internet of Things
IP	Internet Protocol
IPv6	Internet Protocol version 6
IRTF	Internet Research Task Force
LLN	Low power and Lossy Networks
LPWAN	Low-Power Wide-Area Network
MAC	Media Access Control
MQTT	Message Queuing Telemetry Transport
MTU	Maximum Transmission Unit
NDN	Named Data Networking
NPSN	Named Publish Subscribe Networking
OPEX	Operational Expenditure
PIT	Pending Interest Table
PURSUIT	Publish-Subscribe Internet Technology

RPL	IPv6 Routing Protocol for LLNs
SDP	Service Discovery Protocol
SSL	Secure Sockets Layer
TCP	Transmission Control Protocol
UDP	User Datagram Protocol
UPnP	Universal Plug and Play
URI	Uniform Resource Identifiers
UUID	Universally Unique Identifier

References

1. G. Xylomenos, C. N. Ververidis, V. A. Siris, N. Fotiou, C. Tsilopoulos, X. Vasilakos, K. V. Katsaros, and G. C. Polyzos. A survey of information-centric networking research. *IEEE Communications Surveys Tutorials*, 16(2):1024–1049, 2014.

2. K. Pentikousis, B. Ohlman, D. Corujo, G. Boggia, G. Tyson, E. Davies, A. Molinaro, and S. Eum. Information-Centric Networking: Baseline Scenarios. RFC 7476 (Informational), March 2015.

3. D. Trossen, M. Sarela, and K. Sollins. Arguments for an information-centric internetworking architecture. *SIGCOMM Computing Communication Review*, 40(2):26–33, 2010.

4. A. Ghodsi, S. Shenker, T. Koponen, A. Singla, B. Raghavan, and J. Wilcox. Information-centric networking: Seeing the forest for the trees. In *Proceedings of the 10th ACM Workshop on Hot Topics in Networks*, HotNets-X, pages 1:1–1:6, New York, 2011. ACM.

5. D. Perino and M. Varvello. A reality check for content centric networking. In *Proceedings of the ACM SIGCOMM Workshop on Information-centric Networking*, ICN'11, pages 44–49, New York, 2011. ACM.

6. D. Miorandi, S. Sicari, F. De Pellegrini, and I. Chlamtac. Internet of things: Vision, applications and research challenges. *Ad Hoc Networks*, 10(7):1497–1516, 2012.

7. D. Trossen, M. J. Reed, J. Riihijärvi, M. Georgiades, N. Fotiou, and G. Xylomenos. IP over ICN: The better IP? In 2015 *European Conference on Networks and Communications (EuCNC)*, pages 413–417, Bologna, Italy, June 2015.

8. Y. Zhang, D. Raychadhuri, L. Alfredo Grieco, E. Baccelli, J. Burke, R. Ravindran, G. Wang, A. Lindren, B. Ahlgren, and O. Schelen. *Requirements and Challenges for IoT over ICN*. Internet-Draft draft-zhang-icnrg-icniot-requirements-01, Internet Engineering Task Force, April 2016.

9. M. Amadeo, C. Campolo, J. Quevedo, D. Corujo, A. Molinaro, A. Iera, R. L. Aguiar, and A. V. Vasilakos. Information-centric networking for the internet of things: Challenges and opportunities. *IEEE Network*, 30(2):92–100, 2016.

10. B. Ahlgren, C. Dannewitz, C. Imbrenda, D. Kutscher, and B. Ohlman. A survey of information-centric networking. *IEEE Communications Magazine*, 50(7):26–36, 2012.
11. A. Rao, O. Schelén, and A. Lindgren. Performance implications for IoT over information centric networks. In *Proceedings of the Eleventh ACM Workshop on Challenged Networks*, CHANTS'16, pages 57–62, New York, 2016. ACM.
12. J. Quevedo, D. Corujo, and R. Aguiar. A case for ICN usage in IoT environments. In *Global Communications Conference (GLOBECOM), 2014 IEEE*, pages 2770–2775, December 2014.
13. S.-E. Elayoubi and J. Roberts. Performance and cost effectiveness of caching in mobile access networks. In *Proceedings of the 2nd ACM Conference on Information-Centric Networking*, ACM-ICN'15, pages 79–88, New York, 2015. ACM.
14. J. Quevedo, M. Antunes, D. Corujo, D. Gomes, and R. L. Aguiar. On the application of contextual IoT service discovery in information centric networks. *Computer Communications*, 89:117–127, 2016.
15. J. Suarez, J. Quevedo, I. Vidal, D. Corujo, J. Garcia-Reinoso, and R. L. Aguiar. A secure IoT management architecture based on information-centric networking. *Journal of Network and Computer Applications*, 63:190–204, 2016.
16. L. Zhang, A. Afanasyev, J. Burke, V. Jacobson, P. Crowley, C. Papadopoulos, L. Wang, et al. Named data networking. *ACM SIGCOMM Computer Communication Review*, 44(3):66–73, 2014.
17. W. Shang, A. Bannis, T. Liang, Z. Wang, Y. Yu, A. Afanasyev, J. Thompson, J. Burke, B. Zhang, and L. Zhang. Named data networking of things (invited paper). In *2016 IEEE First International Conference on Internet-of-Things Design and Implementation (IoTDI)*, pages 117–128, April 2016.
18. N.-T. Dinh and Y. Kim. Potential of information-centric wireless sensor and actor networking. In *Computing, Management and Telecommunications (ComManTel), 2013 International Conference on*, pages 163–168. IEEE, 2013.
19. Z. Ren, M.A. Hail, and H. Hellbruck. CCN-WSN: A lightweight, flexible content-centric networking protocol for wireless sensor networks. In *Intelligent Sensors, Sensor Networks and Information Processing, 2013 IEEE Eighth International Conference on*, pages 123–128, 2013.
20. J. Burke, P. Gasti, N. Nathan, and G. Tsudik. Securing instrumented environments over content-centric networking: The case of lighting control and NDN. In *IEEE NOMEN Workshop*, Turin, Italy, 2013.
21. J. Burke, P. Gasti, N. Nathan, and G. Tsudik. Secure sensing over named data networking. In *IEEE Network Computing and Applications (NCA)*, Cambridge, MA, pages 175–180, 2014.
22. M. Amadeo, C. Campolo, and A. Molinaro. Multi-source data retrieval in IoT via named data networking. In *Proceedings of the 1st International Conference on Information-centric Networking*, INC'14, pages 67–76, New York, 2014. ACM.

23. D. Corujo, R. L. Aguiar, I. Vidal, J. García-Reinoso, and K. Pentikousis. Research challenges towards a managed information-centric network of things. In *European Conference on Networks and Communications, EuCNC 2014, Bologna, Italy, June 23–26, 2014*, pages 1–5, 2014.

24. J. Quevedo, D. Corujo, and R. Aguiar. Consumer driven information freshness approach for content centric networking. In *Computer Communications Workshops (INFOCOM WKSHPS), 2014 IEEE Conference on*, pages 482–487, April 2014.

25. J. Quevedo, R. Ferreira, C. Guimarães, R. L. Aguiar, and D. Corujo. Internet of Things discovery in interoperable Information Centric and IP networks. *Internet Technology Letters*, 1(1), 2018

26. E. Baccelli, C. Mehlis, O. Hahm, T. C. Schmidt, and M. Wählisch. Information Centric networking in the IoT: Experiments with NDN in the wild. In *1st ACM Conference on Information-Centric Networking (ICN-2014)*, New York, pages 77–86, 2014.

27. MQTT Version 3.1.1. Edited by Andrew Banks and Rahul Gupta. 29 October 2014. OASIS Standard. http://docs.oasis-open.org/mqtt/mqtt/v3.1.1/os/mqtt-v3.1.1-os.html. Latest version: http://docs.oasis-open.org/mqtt/mqtt/v3.1.1/mqtt-v3.1.1.html.

28. Y. Yu, A. Afanasyev, D. Clark, V. Jacobson, L. Zhang et al. Schematizing trust in named data networking. In *Proceedings of the 2nd International Conference on Information-Centric Networking*, pages 177–186, 2015. ACM.

29. J. Ren, K. Lu, S. Wang, X. Wang, S. Xu, L. Li, and S. Liu. VICN: A versatile deployment framework for information-centric networks. *IEEE Network*, 28(3):26–34, 2014.

30. S. Wang, J. Bi, J. Wu, X. Yang, and L. Fan. On adapting HTTP protocol to content centric networking. In *Proceedings of the 7th International Conference on Future Internet Technologies*, CFI'12, pages 1–6, New York, 2012. ACM.

31. Ilya Moiseenko and Dave Oran. TCP/ICN: Carrying TCP over content centric and named data networks. In *Proceedings of the 3rd ACM Conference on Information-Centric Networking*, ACM-ICN'16, pages 112–121, New York, 2016. ACM.

32. A. Afanasyev, I. Moiseenko, and L. Zhang. ndnSIM: NDN simulator for NS-3. Technical Report NDN-0005, NDN, October 2012. http://named-data.net/wp-content/uploads/TRndnsim.pdf

33. S. Mastorakis, A. Afanasyev, I. Moiseenko, and L. Zhang. ndnSIM 2: An updated NDN simulator for NS-3. Technical Report NDN-0028, Revision 2, NDN, November 2016. http://named-data.net/techreport/ndn-0028-1-ndnsim-v2.pdf

34. S. Cheshire and M. Krochmal. Multicast DNS. RFC 6762, February 2013. https://www.rfc-editor.org/info/rfc6762

35. Digital Living Network Alliance: DLNA Networked Device Interoperability Guidelines Expanded. http://www.dlna.org.

36. R. Ravindran, T. Biswas, X. Zhang, A. Chakraborti, and G. Wang. Information-centric networking based homenet. In *Integrated Network Management (IM 2013), 2013 IFIP/IEEE International Symposium on*, pages 1102–1108, May 2013.

37. G. Scott. CCNx 1.0 Simple service discovery. Technical report, Computing Science Laboratory, Palo Alto Research Center, 2014.
38. R. Ferreira, R. Aguiar, and A. Matos. Recognizing entities across protocols with unified UUID discovery and asymmetric keys. In *Global Communications Conference (GLOBECOM), 2013 IEEE*, 2013.
39. P. Leach, M. Mealling, and R. Salz. A Universally Unique IDentifier (UUID) URN Namespace. RFC 4122 (Proposed Standard), July 2005.
40. R. Moskowitz, P. Nikander, P. Jokela, and T. Henderson. Host identity protocol. RFC 5201 (Experimental), April 2008. Updated by RFC 6253.
41. A. Delphinanto, J. J. Lukkien, A. M. J. Koonen, F. T H Den Hartog, A. J. P. S. Madureira, I. G. M. M. Niemegeers, and F. Selgert. Architecture of a bi-directional Bluetooth-UPnP proxy. In *Consumer Communications and Networking Conference, 2007. CCNC 2007. 4th IEEE*, pages 34–38, January 2007.
42. B. Villaverde, R. Alberola, A. Jara, S. Fedor, S. Das, and D. Pesch. Service discovery protocols for constrained machine-to-machine communications. *Communications Surveys Tutorials, IEEE*, pages 1–20, 2013.
43. A. Wils, F. Matthijs, Y. Berbers, T. Holvoet, and K. De Vlaminck. Device discovery via residential gateways. *Consumer Electronics, IEEE Transactions on*, 48(3):478–483, August 2002.
44. S.-H. Kim, J.-S. Kang, H. S. Park, D. Kim, and Y.-J. Kim. UPnP-ZigBee internet working architecture mirroring a multi-hop ZigBee network topology. *Consumer Electronics, IEEE Transactions on*, 55(3):1286–1294, August 2009.
45. M. B. Krishna and D. Goyal. Mobile middleware application and services for real-time water pump system. In 2015 *2nd International Conference on Signal Processing and Integrated Networks (SPIN)*, pages 411–417, February 2015.
46. Z. Shelby, K. Hartke, and C. Bormann. The constrained application protocol (CoAP). RFC 7252 (Proposed Standard), June 2014. Updated by RFC 7959.
47. D. Kutscher, S. Eum, K. Pentikousis, I. Psaras, D. Corujo, D. Saucez, T. C. Schmidt, and M. Whlisch. Information-centric networking (ICN) research challenges. RFC 7927, July 2016.
48. J. Quevedo, C. Guimarães, R. Ferreira, D. Corujo, and R. L. Aguiar. ICN as network infrastructure for multi-sensory devices: Local domain service discovery for ICN-based IoT environments. *Wireless Personal Communications*, 95(1):7–26, 2017.
49. R. L. Aguiar. Some comments on hourglasses. *SIGCOMM Computer Communication Review*, 38(5):69–72, 2008.
50. L. Muscariello, G. Carofiglio, and M. Gallo. Bandwidth and storage sharing performance in information centric networking. In *Proceedings of the ACM SIGCOMM Workshop on Information-centric Networking*, ICN'11, pages 26–31, New York, 2011. ACM.

4

INFORMATION-CENTRIC NETWORKING FUTURE INTERNET VIDEO DELIVERY

JORDI ORTIZ, PEDRO MARTINEZ-JULIA AND ANTONIO SKARMETA

Contents

4.1 Introduction

Nowadays content delivery, with an emphasis on video streaming, is the major source of bandwidth consumption on the Internet, so efficient and effective content distribution is a key aspect to deploy bandwidth demanding services at large scales. As a solution, Content Delivery Networks (CDNs) are being deployed and offered to content providers and carriers as cost saving solutions, but they add undesired complexity to end-to-end operations and content state management, becoming too closed and hard to adapt to new workflows. Those solutions are also limited by the strict environment established by underlying architectures, which are typically tied to Internet Protocol (IP) and conducting to the ossification of networks.

Guided by these limitations of the current network architectures, the major proposals towards the Future Internet (FI) have focused on providing an evolutive or disruptive path to deliver the best service to final users. On the one hand, we have the cases of Location/ID Separation Protocol (LISP) [1,2], Host Identity Protocol (HIP) [3], Heterogeneity Inclusion and Mobility Adaptation through Locator ID Separation (HIMALIS) [4], and Mobile Oriented Future Internet (MOFI) [5], which provide different solutions to separate identifiers and locators in the network in order to provide enhanced scalability and security.

On the other hand, in order to resolve part of the limitations exposed by CDNs and thus break the ossification of networks in content distribution, the Information-Centric Networking (ICN) paradigm has proliferated to exploit the possibilities that intermediate network elements have to achieve efficient and effective content delivery. Within ICN we find some outstanding solutions, such as NDN/CCN [6], PURSUIT [7], and Network of Information (NetInf) [8], which provide mechanisms to change the way content is represented in the network traffic and intermediate elements.

Although all architecture proposals have represented a renovation of the network to provide specific solutions to specific problems, most of them are clean-slate, so they require a jettisoning of the current infrastructures in order to support the building of their solutions. However, there is a distinctive approach that opens evolutive paths and allows the introduction of new mechanisms to current networks. It is the SDN initiative [9].

With the support of SDN, some mechanisms can be easily adopted in current networks, as we describe throughout this chapter. This implies that network endpoints can be kept unchanged but the specific protocols and architectures underneath their communications will be adapted to achieve more efficient operations. As also discussed in this chapter, SDN can be used efficiently and effectively to address content delivery in current networks without requiring clients or servers to change the way they are attached to the network (e.g. IP network).

In addition, devices' heterogeneity today is seen in the IoT concept and the Future Internet initiative. How the multimedia content produced and consumed by these devices and their specifically designed networks is transmitted is a field in which ICN and SDN can have some relevance in a *near future* which is becoming every day more *the present*.

The remainder of this chapter is organized as follows. First, in Section 4.3, we introduce the most representative solutions related to FI and ICN, while contextualizing the work we present in this chapter. Then, we describe our proposal on ICN applied to IoT in Section 4.4, followed by an evaluation of the proposal in Section 4.5. The Future Research Section 4.6 aims to introduce open research fields in ICN and FI. Finally the chapter is concluded in Section 4.7.

4.2 Background

The Internet was begun with the objective of establishing a simple solution to allow remote computers to exchange messages without monopolizing current communication infrastructures and to provide the ability to recover from underlying failures. However, such infrastructure has become the center of our digital lives and, inevitably, it has shown several problems. The definition of new architectures has been supported, as have alternatives to overcome such limitations and allow the network to advance and thus improve the service offered to users.

4.2.1 Separation of Identifiers and Locators

Instead of fixing the problems with complementary mechanisms, there is some consensus in the research community that some problems related to end-to-end communications may be resolved by separating the location and identifier of a network node. This approach is followed by two outstanding architectures, HIP [3] and LISP [1,2], as well as other derivatives and totally new proposals.

The HIP introduces cryptographic host identifiers forming a new global name space as a new intermediate layer between the IP and transport layers. It decouples the endpoint identifier and locator, enabling the transport on host identifiers and routing on IP addressing that serve as pure locators. Although this seems a good solution, it presents many problems because of the intrinsic meaning of identifiers and, in general, its weak solution to all requested capabilities for the Future Internet.

On the other hand, the LISP is a routing-based solution using map-and-encap at border routers. The upstream IP address of border routers is used as Routing Locator (RLOC) of hosts residing in the local domain to perform inter-domain routing, while intra-domain routing is performed using conventional IP addresses, which are also referred to as Endpoint Identifiers (EIDs). These EIDs can potentially be associated with a group of RLOCs to support multi-homing. LISP tunnels data packets between source and destination RLOCs using the LISP database, which contains the RLOC/EID relation for each local domain to interpret packets. It also uses an on-demand cache to select the destination RLOC for sending packets towards specific destinations. As the main drawbacks exposed by this architecture, we highlight the need to maintain a LISP database, because it raises scalability concerns, and that the creation of an on-demand cache to route packets among different domains is not clearly defined. As a HIP derivative, the BLIND architecture [10] enhances HIP with security and identity protection, as well as location privacy, by introducing new forwarding agents. Even though this is an interesting architecture from a security point of view, it inherits the same problems that HIP has.

Also similar to HIP, the Routing Architecture for the Next Generation Internet (RANGI) [11] introduces a new layer called *Node Identity Internetworking Architecture (NodeID)*, which is used to transport related communications instead of IP addresses. RANGI adopts a hierarchical structure of host identifies employing special IPv6 addresses. The main benefit of RANGI is scalability, and its main drawback is the complexity related to fundamental modifications of the IP addressing scheme and the Domain Name System (DNS) entry to store the required mappings.

The NodeID [12] proposal from the EU FP6 *Ambient Networks* project is also similar to HIP and LISP and introduces an architecture based on separation of locator, address, and administrative domains. Nevertheless, it is a network layer solution based on a locator/identifier split approach, and thus does not address session or identity management at all. Also, this architecture suffers from scalability problems related to inter-domain routing, a process based on routing tags, which contain domain information about the location of a concrete node identifier.

Defined within the AKARI project, which is being carried out by the National Institute of Information and Communication Technology of Japan (NICT), we find the HIMALIS architecture [4]. It proposes a complete architecture that shares features with HIP and LISP but targets a new identification scheme, different from IP and capable of supporting sensor networks and the IoT. The major benefits of HIMALIS reside in the simple targets of heterogeneity and mobility. It provides the necessary functional blocks to complement the current Internet architecture and allow it to evolve into a more functional network. However, there are still functions not covered by this architecture, especially those related to addressing the identities of network elements and managing communications according to them.

The MOFI [5] proposes an architecture that considers, from the beginning, that network elements (hosts) can move across networks while also considering the existence of current IP networks as a backbone for communicating separated edge networks. In contrast, it proposes new control and data planes for those edge networks to optimize mobility. The simplicity of this architecture and the organization of the functions provided in separated functional blocks (components) make it very interesting vis a vis adopting new Internet functions. However, it does not address security or discovery problems.

In a less disruptive approach by employing DNS for name to identifier and to address translation, the ILNP [13] architecture faces the id/loc split. It also suggests the inclusion of the identifiers as part of IPv6 addresses. The approach is too tied to DNS which has a slow convergence time suffering in the dynamicity side, a undesirable side effect for FI. It is being evaluated nevertheless by the Internet Research Task Force(IRTF)[14].

4.2.2 Integrated Content Delivery

The vast majority of data exchange in the Internet are produced between a provider and a consumer, therefore it can be cataloged as information-centric communications, in the sense that they are not node oriented like the actual addressing system. The ICN approach breaks with the node centric philosophy of the Internet in a disruptive yet well-accepted way for the future Internet. Unlike present end-to-end communication approach, information centric communication is the focus of the research community. These communications do not connect nodes with links and processes any more but connect consumers with producers in a decoupled way.

The NetInf [8] paradigm, which extends the concept of identifier/locator separation by introducing another level of indirection and decoupling self-certifiable objects from their storage locations, was introduced and motivated by the EU-funded projects 4WARD [15] and SAIL [16]. The paradigm distinguishes between information and data, where the former represented with Information

Objects (IO) is independent from the actual encoding that is managed by the Data Objects (DO). This way semantic relations between different sources can be modeled as well as the relations between copies of the same content. Users can, thanks to the self-certifiability of the objects, access the closer copy of the element being addressed, optimizing the distribution chain.

Another approach to ICN is the publish/subscribe scheme born from the also EU-funded, Publish-Subscribe Internet Routing Paradigm (PSIRP) [17] and Publish-Subscribe Internet Technology(PURSUIT) [7]. In this case, the consumer subscribes to the desired information while providers publish it. The approach adds different identifications schemes for information, subscribers, publishers, and intermediate nodes. A less clean slate approach, since it relies on the actual Internet meaning IP routing, replaces the typical DNS name resolution with self-certifying flat names and a name-based anycast primitive. The Data-Oriented Network Architecture (DONA) [18] does not certify the content itself but the publishers, labeling the data produced by them. Unfortunately, this approach does not allow dynamic data generation since it must be first registered in the trusted resolution handlers (RHs). Security is addressed within DONA by validating content and providers.

Similarly to the pub/sub approach of PSIRP/PURSUIT and adding the closer copy distribution concept of NetInf, the Content-Centric Networking (CCN) [6] approach proposes the distribution of content-centric elements on every network element, let it be a switch or a router. In this approach, the consumers provide the network with their interests, and the network transparently solves the content location resolution and transport. One of the main advantages of CCN is its similarity to CDN as a globally distributed cache. Although its extensibility [19] and support for different traffic types, it has no accounting for identifying the communication peers, therefore lacks end-point security or privacy per design.

4.2.3 Video Streaming Requirements in the Internet

4.2.3.1 Video on Demand, Live Video Streaming, and Dynamic Adaptive Streaming In general, any video file can be uploaded to a server with any file transfer protocol such as FTP and be downloaded by similar means, HTTP among others. Nevertheless, downloading a video file from a content server as it is and reproducing it afterwards is not considered video streaming. A video streaming or video on demand service in addition to providing the data to the client, controls the delivery rate with real-time playing capabilities. In addition, it is considered that a streaming service should provide video cassette recorder (VCR) capabilities, meaning that the stream can be paused, rewound, or fast forwarded (in case of non-live content). In the case of live video streaming, the content needs to be encapsulated onto the transport mechanism on the fly. The major difference between video on demand and live streaming is the absence of a prearranged end time [20].

Both video on demand and live streaming depend on the amount of available bandwidth to provide a real-time service. If the available bandwidth is cut below the video bit rate being streamed, the quality of experience (QoE) perceived by the user is reduced. To mitigate or even eliminate the drawbacks caused by network shortage, adaptation is performed. In the case of live streaming only transcoding, i.e, decoding and reencoding of the video source with different parameters, is possible unless a scalable video codec is used. With video on demand the possibility of encoding multiple formats and storing them, either in different files or in a multi-hint container, is also possible. Adaptation might produce other drawbacks related to the switching between bit rates such as reinforced artifacts or video quality degradation and upgrade switch; these changes can produce a loss in the QoE to a user who might prefer a constant reduced quality [21].

4.2.3.2 Evolution of Video Transmission Techniques   Interest in video transmission on the Internet started in the early 1990s [22]. This interest appeared partly as a consequence of the increase in computer processing power. This increase allowed real-time capture, compression, and transmission of video streams. The research in multicast IP and other related datagram transmission techniques also contributed to increase interest in this field. Real-time Transport Protocol (RTP) [23] was introduced by the IETF in the period 1992–1996 and became the sole transmission protocol for multicast conferencing systems. RTP allows not only data transmission but also managed data control. In addition to RTP as a transport protocol, a solution for announcing multicast sessions was introduced, Session Announcement Protocol (SAP) [24]. In addition, to describe the transport streams the Session Description Protocol (SDP) [25] was introduced. In parallel with the multicast conferencing systems the WWW emerged. The increases in bandwidth in addition to the already mentioned increase in computing processing power allowed the inclusion of multimedia, audio and video, content on web pages. The pioneers in this field were Realtime and Quicktime. In 1998 the Real-time Streaming Protocol (RTSP) [26] standard appeared, providing VCR-like control. RTSP relied on already existing protocols and followed a command signaling very similar to the already widespread HTTP. Another protocol worth mentioning is Adobe's Real-Time Messaging Protocol (RTMP) [27].

At the same time video and audio coding techniques evolved. Video coding is considered to have four steps, partition, predict (subtract), transform, and entropy encode; meanwhile video decoding has the reverse four steps, entropy decode, inverse transform, predict (add), and reconstruct. Different partition, prediction, and transform techniques have produced an evolution and a wide variety of video codecs.

The reference in the video coding field was and continues to be the codecs developed by ITU-T Video Coding Experts Group (VCEG) and ISO/IEC Moving Picture Experts Group (MPEG) video codecs, starting from MPEG-1 [28] Part 2 (first released in 1993), which was rapidly continued by MPEG-2 [29] Part 2 (also known as ITU-T H.262, released for the first time in 1995 with the last amendment made in 2013), which became the de facto standard for video storage and transmission. Among its capabilities was to maintain a fixed rate that could be introduced into an Integrated Services Digital Network (ISDN) channel. The next remarkable evolution in video codecs appeared in 2003 (evolving until 2014) with the MPEG-4 [30] Part 10 Advanced Video Coding (AVC), or ITU-T H.264, which is considered today's standard and is integrated in almost every device available. Other alternatives in video coding appeared in this period of time. The VPx family was created by On2 and purchased by Google to make it free for use with open source implementations. The Xiph Theora [31] family, which was born completely open, and the BBC Dirac codec were based in wavelets instead of the typical DCT approach.

In the present the two main video codec standards are High Efficiency Video Coding (HEVC)/H.265/MPEG-H [32] and Google's VP9 [34], both released in 2013. Both codecs focus on reducing the amount of bandwidth requested for a certain quality level at 50% with respect to their predecessors H.264 and VP8, respectively. Other proposals for stereoscopic video and scalable video [34,35] have been produced along the way. Some were related to enhanced visual experiences such as Stereoscopic and Multi View Coding [36] techniques, while others such as Scalability extensions were directed to enhance network and storage adaptability of the codecs.

4.2.3.3 Establishment of HTTP for Streaming With the increase of available bandwidth in the last mile, the need to adjust the stream bit rate to the available bandwidth has become less relevant. It is considered that having twice the bit rate available as bandwidth using Transmission Control Protocol (TCP) (and thus HTTP) is sufficient for achieving good performance with a few seconds' delay [37]. Providing multimedia content over HTTP [38] has some inconveniences related to suboptimal storage, exceeded bandwidth consumption, and complex bit rate logic; nevertheless, the advantages regarding simplicity in deployment, economics, and scalability have surpassed the inconveniences for the industry. In this approach the server becomes a passive element just offering the content and some metadata that allows the client to decide which resources need to be retrieved, in which order, and at what moment in time. The approach is very attractive since any HTTP deployment is easily converted in a video streaming deployment without specific infrastructure. As a counterpart, control of provider bandwidth consumption is lost.

In 2009 Apple introduced their HTTP Live Streaming (HLS) [39] solution, Microsoft proposed the Microsoft Smooth Streaming, Adobe proposed the HTTP Dynamic Streaming, and finally on the standardization bodies we can find the Adaptive HTTP Streaming

(AHS) from the 3GPP and the MPEG DASH [40–42]. Each of these systems is based on its own manifests and file formats. The Apple system uses m3u playlist and video must be encoded with MPEG2-TS. Microsoft requires a server manifest and a client manifest file and defines a smooth streaming format based on the ISO Base media File Format [43]. Adobe makes use of their own Flash media manifest and F4F file format (also based in ISO Base Media File Format). The 3GPP proposal (AHS) defines a Media Presentation Description (MPD) acting as manifest file and also extends the ISO Base Media File Format for storing the media. Finally MPEG proposal (DASH) adopted the 3GPP's MPD description file as a starting point while media segments must be compatible with the ISO Base Media File Format. As can be easily seen there are two common factors in each approach: the definition of a manifest/session file and compliance with the ISO Base Media File Format.

These HTTP-based streaming systems can take advantage of any existing and/or incoming enhanced distribution mechanisms designed for standard HTTP content including caching systems, proxies, load balancers, and of course any URL-based enhancement.

4.2.4 Internet of Things (IoT)

In recent years, machine-to-machine networking, mainly represented by the so-called IoT, has been a trending topic in FI-focused research. Things call for *smart objects* to be integrated in communication networks of today and the future. Heterogeneity is one of the major characteristics of the IoT. These *things* are usually small in terms of constrained memory, energy, and CPU power resources, or even networking capabilities. How these objects should be exposed to the network and how they should offer basic services is also a field of interest among researchers.

As a consequence, in terms of communications, specific protocols and standards have been defined by the standardization bodies, therefore defining a new protocol stack. At physical and link layers of the OSI, model 802.15.4 [44] is used, which is oriented to low-energy communications, limiting what can be done on the upper layers. 6LoWPAN [45] is used to allow IPv6 communications on top of 802.15.4 that allows at most 102 bytes' (127 bytes less the 25 bytes from the MAC layer header) payload per packet in its last revision [46]; therefore, 6LoWPAN provides packet fragmentation and reassembly. Characteristics related to IoT communications have their precedent in Wireless Sensor Networks but in a more isolated environment; meanwhile, new standards try to integrate the networks with existing Internet standards [47]. Above 6LowPAN the Routing Protocol for Low Power and Lossy Networks (RPL) [48] and the Constrained Application Protocol (CoAP) [49] are defined but they are completely out of the scope of this chapter.

4.3 Architecture

In this section we introduce an architecture that faces the distribution of IoT content following the Information-Centric Networking initiative by the means of SDN.

Information, and in particular video, produced or consumed in an IoT network has certain characteristics related to the nature of the devices present in such networks. The low bandwidth provided by such networks contrasts with the number of devices connected to them. In addition, the nature of IoT devices, which often offer temporary and frequently updated information, implies the need of in-network storage and caching.

We envision a system that encapsulates video produced/consumed in an IoT network into DASH streams to ease the distribution; meanwhile, the CDNaaS is introduced to allow the deployment of caches with a certain level of granularity. Moreover, the CDNaaS potentially allows the easy aggregation of information by means of creating multiple CDNs by device type or user subscription among other possibilities. Thanks to SDN and by means of delayed binding the system is able to inspect the HTTP headers and transparently redirect the HTTP requests to the desired cache based on the URL and not in the communication endpoints. Unlike other ICN alternatives, there is no need for specific software installation per routing node, similarly to CCN. The proxy is introduced in order to mitigate the impact of delayed binding on the CDN/SDN controller, also allowing the distribution of the load on multiple proxies for the same CDN.

Figure 4.1 shows how caches and proxies could be deployed in different positions depending on the role and/or in the sources and sinks of information. As an example, traffic information can be provided only to traffic agents, and dedicated caches are deployed near the agents; on the other hand, weather information is cached in the border near the retrieval point, near premium subscribers as well as near other SPs (to save bandwidth). Let us clarify that although we consider our proposal to be a user-oriented solution, network operators could profit

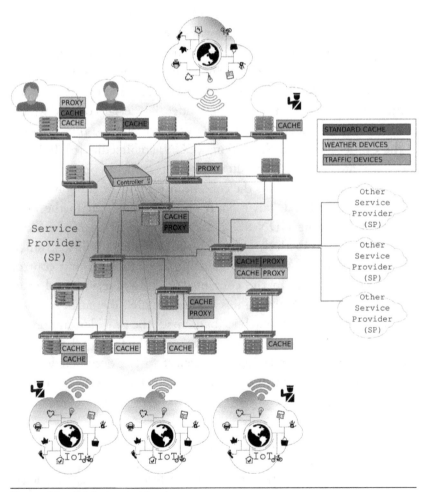

Figure 4.1   Architecture envisioned.

from it by instantiating independent caching deployments based on network usage or other parameters.

This section is split in two parts, the first related to the transport of Internet of Things video onto ICN and the second introducing the concept of CDNaaS by the means of SDN networking and ICN oriented. Both approaches for IoT and SDN are envisioned as the present and the future of computer networks. Meanwhile, video delivery is already a topic of interest due to its impact on current network utilization. Although both parts are evaluated separately they have a common key technology, HTTP/DASH, unifying them.

We propose involving the promising DASH technology within the IoT environment to ease and enhance the video experience on it. In addition, and to ease the transmission of DASH on top of the constrained IoT networks, we propose the introduction of CCN as an abstraction layer for DASH which will ease the transmission and caching of the DASH video content and the location by the network entities. In relation to the fact that video is special content which should be treated carefully by the network, we propose the adoption of the Content-Centric Networking (CCN) architecture [6]. In brief, the CCN architecture is an alternative approach to communication networks where the focus is the content instead of the endpoints.

On the other hand, our second proposal is related to the possibility of arranging CDNaaS thanks to the adoption of the SDN [50,51] paradigm. With this proposal the customer is able to arrange his own caching architecture, reducing the influence and control of network operators in the content distribution service and even obliterating the CDN companies business model. By means of SDN the old host-to-host approach is abandoned, the providers announce the content to be available in the CDN, and the CDN takes care of requested content by relaying the request to the designated cache without provider intervention in subsequent requests, thus becoming an ICN solution.

4.3.1 IoT Video Transport Impact on CCN

IoT devices are inherently constrained in terms of battery capacity. The battery restriction has side effects in the limitations produced in the delivered computing power as well as in the wireless connectivity. The former restriction affects, in terms of video coding/decoding, not only the OPS[1] that the devices can execute but also the amount of memory needed to store the decoded and the coded video no matter if the device is encoding or decoding. The latter implies the limitation of the bit rate and the packet size.

Considering the well-known 6LoWPAN [52,53] architecture and the top speed specified by the IEEE 802.15.4 [54], a video that is going to be directly transmitted on top of this kind of network must not exceed its design specifications. The packet maximum data payload is to be limited to less than 100 bytes, while the bit rate must stay below 250 kbps, which would be a network's characteristics under optimal conditions; in the real world these values (mainly the bit rate) would be lower. In terms of codecs, one can assume at least the support for H264/AVC [55] (also known as AVC), which has been considered the de facto standard for video and which was designed for accommodating transmissions starting from 64 kbps [56], the size of a T1 channel and ISDN increments.

The memory restrictions, taking into account that the thing also needs an operating system and some libraries and assuming that there is no specific memory or processing unit for video, imply a limitation assuming less than 1 kB ram. Although for frame size bigger is better, the codec mechanisms to reduce the output bit rate are also taken into account. In that respect, prediction is a key mechanism to produce such reduction, meaning that at least two frames need to be present in raw format into the memory. Considering that a Common Intermediate Format (CIF) (352 × 288 pixels) raw image needs approximately 297 bytes, two frames in this format would already take up half the device's memory, leaving the other half to store the coded version. The AVC baseline profile is to be used to avoid the inclusion of B frames which require the storage of at least two raw frames. We foresee also a reduction from the typical 25 frames per second (FPS) to a more likely 5 FPS.

Taking into account the above mentioned restrictions, some video streams were pre-encoded using the H264/AVC with the baseline

---

[1] Operations per second

profile, limiting the slice size to 81 bytes so that packet splitting and aggregation at the network level are not needed with 250 kbps as the peak level bit rate. The produced streams have 38, 58, and 79 kbps average bit rate, considerably lower than the maximum specified previously since we have taken into account that the effective bandwidth is usually considerably lower than the theoretical maximum and adaptation will be required.

We have also decided to restrict the encoding to intraframes so that the thing needs only one decoded frame memory size. But what is more important is removing the complexity introduced by prediction, motion estimation, and compensation procedures. Furthermore, the difference in bit rates was achieved by the use of a fixed quantization parameter (QP). The video used is a concatenation of the well-known *bridge* and *bride_far*[2] streams, which produce an 8200 frame or a 164 second video stream at 5 FPS.

The video streams have been encapsulated in DASH [40]. DASH is considered the video transmission killer technology. The all-HTTP philosophy offers a well-known technology for the industry, providing not only a high degree of confidence but also deployment procedures that are already evolved and an architecture deployed worldwide. CDNs are also part of these inherited advantages. Another key feature for DASH is the dynamic bit rate adaptation which is completely managed by the client, avoiding the need for modifications on the server side. Therefore, we have encapsulated the multiple bit rates available for the same stream in the same container which allows receiving multiple descriptions in the same MPD [40]. The reasons a client might decide to switch the media version to be received can vary from the simplest and more straightforward bandwidth or hardware limitations (e.g. screen size) to more complex and elaborate ones such as those that are user driven [57]. For this evaluation a single bit rate is forced to ease the study of the network and allow the production of objective video analysis such as PSNR values. We want to demonstrate that the changes in perceived quality are produced due to network characteristics and not as an effect of particular software bandwidth estimation and adaptation decisions. As a consequence of the restrictions imposed

---

[2] http://media.xiph.org/video/derf/

in IoT-constrained scenarios, the evaluation of DASH over CCN shows, as expected, reduced performance in comparison with typical IP-to-IP and host-to-host scenarios.

While encapsulating the video streams the DASH video chunk size was set to the minimum allowed by the encapsulation software [58] which corresponded to one frame – one chunk, thus taking advantage of the ability of CCN to transmit very small fragments.

To fulfil the requirements the "libdash" [59] implementation and DASHEncoder from ITEC-UNIKLU was used with slight modifications to accept the bit stream generated with JSVM 9.19 [60] as input.

As part of the experimentation, HTTP-to-CCN (client side) and HTTP-to-CCN (server side) adapters were implemented despite having those that are part of CCNx available. The latter were evaluated and were considered unsatisfactory in terms of stability and user experience. The implemented adapters make use of PYCCN [61] library and are therefore PYTHON based. This library is provided by UCLA and offers Python bindings for the C version of the library which speeds up the development by means of having an OOP-style interface to CCNx.

At the end, DASH HTTP chunks can be much bigger than the MTU of the underlying network while they are fragmented into very small content objects and transmitted by CCN to the *thing* (DASH client).

In order to evaluate the feasibility of transporting IoT-compatible video data by means of HTTP streaming over an ICN network, a deployment on top of PlanetLab [62] was done. PlanetLab is a well-known global infrastructure for network research in which the CCN network composed of multiple CCN routers as well as the CCN/HTTP Adapter Server and HTTP/CCN Adapter client, collocated with the HTTP/CCN Adapter client, is deployed with the DASH client itself. Part of the GAIA testbed [63] was also employed to deploy the Cherokee HTTP server[3] in charge of serving the MPD files as well as the related DASH video chunks. The deployment can be seen in Figure 4.2.

---

[3] http://www.cherokee-project.com

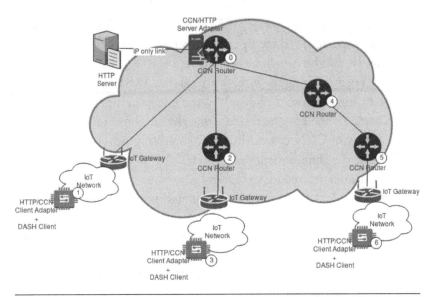

Figure 4.2   Overview of the experimentation architecture. It shows how the CCN network we used in our experiments was built, including CCN routers and CCN/HTTP Adapter Server and client instances.

To ensure repeatability, experiment control was performed via the Network Experimentation Programming Interface (NEPI) [64] framework. The framework receives as input the nodes involved in the experiment, the applications to be run on top of them, as well as other topology details. The outputs to be obtained to be later analyzed are also specified. NEPI takes care not only of experiment deployment but also the reservation of the resources in an experimentation architecture like PlanetLab where such an experiment is going to be deployed. NEPI also deploys the applications on top of the scenario and synchronizes the experiment execution, ensuring the same timing is achieved among experiment runs, of which it also takes care. Finally NEPI gather results and provide them to the experimenter. The combination of PlanetLab and NEPI ensures repeatability and results validity. In this case the capability of NEPI to create IP tunnels (as can be seen in Figure 4.3) is used to create an overlay on top of the Internet, interconnecting the nodes of the PlanetLab infrastructure to be used as part of the experiment. The Round Trip Time (RTT) of every hop is ensured to be above 30 ms due the nature and dispersion of PlanetLab. The high RTT and node spread location allow experimentation results to approach real-world results.

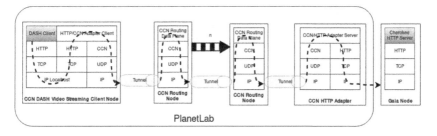

Figure 4.3 CCN entities and relations.

The CCN/DASH Video Streaming Client node contains both the DASH client and the aforementioned HTTP/CCN Adapter Client so that the interaction between the two is negligible and ensures that the network evaluation represents only the CCN communication. The CCN/HTTP Adapter Server acts also as a CCN Router, allowing in-network caching for the rest of the nodes. To evaluate the effect of CCN routing, the streaming nodes were situated one, two, and three CCN jumps away from the server in what can be seen as different network domains (Figure 4.2). The smaller IoT networks represent the constrained network which is simulated by manipulating the packet size of the CCN exchanges.

The experimentation was directed to two different scenarios, one in which the video chunks are retrieved without limiting the amount of time needed, which could probably lead to either video playing stop or chunks being discarded for arriving late, and a second scenario in which the timing of the transported chunk is taken into account to timeout the request (corresponding to a live streaming process), meaning that bizarre effects or jumps in the video stream could appear.

Figure 4.4 is a sequence diagram that details the interactions between the DASH client, the adapters, the CCN routers, and the HTTP server. The figure shows the sequence for the first retrieval of an MPD file, which in turn is the first file from a DASH stream to be retrieved, and a retrieval from the same content from another client in another point of the network and how this second request produces a cache hit in the CCN router nearer to the content source, i.e, the HTTP server. The figure is also valid also for video chunk retrieval. Taking into account the deployment in Figure reffig:architecture, multiple requests for the same content will lead just to one content

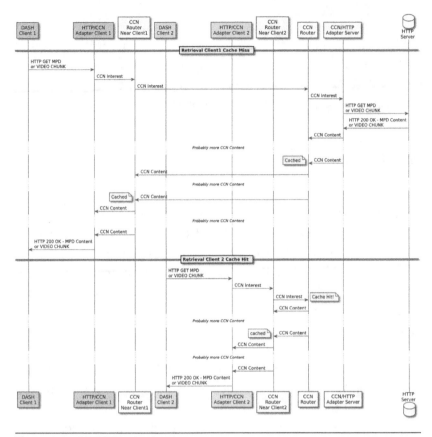

Figure 4.4   DASH/CCN network exchange sequence.

retrieval from the content provider while the rest of the requests will be solved by the CCN/HTTP Adapter Server cache unless content is removed due a cache replacement policy.

4.3.2 Leveraging Software-Defined Networking for Information-Centric Video Distribution

As introduced at the beginning of the Section 4.1, video consumption is fostering innovation in content distribution systems that can profit from the so-called future Internet technologies.

As introduced above, DASH is the acronym for Dynamic Adaptive Streaming over HTTP. DASH [40] relies on the HTTP client server model in addition to eXtensible Markup Language(XML) to provide on top of the Internet efficient and high-quality media delivery. XML is employed to expose the metadata needed to describe technically and index the media to be streamed so that it can be deployed on standard HTTP servers and accesses like any other web page. The complexity on the streaming process is delegated to the client which is in charge of parse the metadata file and download the corresponding media data.

To simplify media transmission, DASH allows the splicing of media files into chunks that are indexed with the MPD file (holding the metadata). Each MPD file may contain different media representations which correspond to different characteristics of the same media source. Those representations allow clients to select the operation point or the media characteristics, such as bitrate or frame size, to be delivered.

The key characteristic of DASH is that the files forming a stream, mainly video chunks and the MPD file, can be stored in a standard HTTP server and be retrieved like any other web page. Likewise, any optimization performed to the HTTP ecosystem, such as caching systems or CDNs, is seamlessly incorporated to the content retrieved via DASH. In addition, DASH also defines profiles for different media streaming approaches like live content, which provides interoperability and adequation to different content provider needs.

On the other hand, a more disruptive approach for media transmission is the one provided by Content-Centric-Networking (CCN) (Introduced in Section 4.2.2) as an alternative to TCP/IP and CDN model.

In this section a Content- and Information-Centric Distribution Network as a service is defined, and extending the XaaS ecosystem by employing SDN-related technologies is introduced. The goal is to provide the means to offer content providers with differentiated caching services provided with enough dynamism to cope with changing user behavior to offer the best quality of experience.

4.3.2.1 CDNaaS Concept and Motivation   Content distribution services are traditionally implemented through a mix of techniques like HTTP redirection, DNS load distribution, anycast routing, and application-specific solutions, among others. As a result, a complex distributed system is in charge of redirecting users' requests to clusters of network caches. Such decisions, i.e. the right cache(s) that must serve the content, are usually made based on the communicating endpoints (IP addresses) involved. Contrary to this, the ICN paradigm advocates for delivering requested resources based on their name and independently from the data transport [14]. This can potentially increase the efficiency and scalability of content distribution, but it typically requires the deployment of state-of-the-art protocols like CCNx [6] as introduced in previous chapters.

On the other hand, in the last few years we have witnessed the rise of SDN and the high momentum they have gained [65]. By means of a logically centralized controller that maintains the global view of the network and exposes a programmatic interface, SDN offers huge opportunities for network programmability, service automation, and simplified management.

Can we leverage SDN to provision CDNaaS which are led by ICN principles? What are the benefits of such an approach? What challenges must be faced? Our claim is that provisioning CDN services with SDN is possible without a considerable loss of performance while increasing the dynamism of caching systems by adopting the ICN principles.

We consider that an enhanced service to be provided by service providers would enable the content provider to arrange its own CDN within the service provider's premises. With that and the adoption of ICN initiatives, the service provider benefits from uplink bandwidth consumption reduction in cases in which the content provider is not within the service provider's network and also market diversification by offering CDN services, clearly differentiated from the competence, thanks to the integration on the network. As a consequence of adopting SDN, the CDN can be easily and dynamically rearranged, and the provider himself could interact with the system. Also the caching mechanism can be modified on demand as can the content to cache assignment algorithms. The producer could potentially select where the data is stored based on the data itself and not only on the consumer. In addition, our system avoids contacting the provider each time some content is requested; it is the network itself that resolves the request. Meanwhile, the client receives content transparently, if desired by the provider, or after any bootstrapping action that could be also configurable, like payment or successful authentication among others.

4.3.2.2 Design   We have designed, implemented, and deployed a proof-of-concept SDN application that realizes CDNs in network providers that feature a fully operational SDN infrastructure with at least one OpenFlow [66] controller. In addition, the operator needs network appliances or (either virtual or physical) general-purpose computers that perform network functions, such as web caching. The system allows a network operator to instantiate multiple CDNs concurrently over the same architecture for different providers. These instances could potentially adopt different caching policies which would be enforced dynamically by the SDN application.

The Network Function Virtualization (NFV) aims to address the problem of network service instantiation delay due to hardware slow release periods and obsolescence by leveraging standard IT virtualization technology. It is envisioned as a complement of SDN in such a way that the term *SDNFV* [67] has been coined. In that line the Management and Orchestration (MANO) stacks are defined and evolved to coordinate and instantiate services on top of the virtualization premises as well as organize the network to support those services. We argue that our defined API would fit in MANO stacks, and some publications [68] have already envisioned the CDN as one of the most interesting services to be virtualized, although they followed a more traditional approach.

Our SDN application has been built on top of the Floodlight controller [69] and exposes a RESTful Northbound API that is exploited by network management tools to control the life-cycle of delivery networks. In our case we have used OpenNaaS [70]. By means of the former API, a CDN can be associated to (at least) a content provider and a set of web caches. The application is in charge of placing and retrieving resources based on their names (Uniform Resource Identifiers [URI] in HTTP requests), borrowing the benefits of ICN without requiring the deployment of new distribution protocols. In addition, having a centralized view of the network facilitates the development of intelligent policies that determine the best placement of each content resource. The application could also perform access control, allowing only selected clients or selected parts of the network to access the content.

In order to implement name-based content placement and retrieval, the SDN application must inspect HTTP flows originated from consumers and targeted at providers within a delivery network. However, the application cannot find out the resource URI until the TCP three-way handshake has completed. This is problematic because the application must direct the flow to the appropriate cache or origin server since the first TCP SYN segment. To overcome such issues, we have implemented a flexible HTTP proxy that performs delayed binding (aka TCP splicing) and provides our SDN application with the name of the requested resource. Details on the message sequence are shown in Figure 4.5 and take into account that multiple proxies could be instantiated and that the decision on which one should be used is taken by the SDN app. For the latter, we have

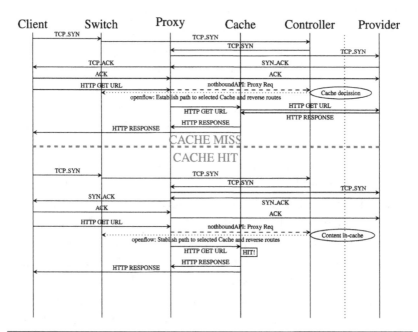

Figure 4.5   Sequence diagram.

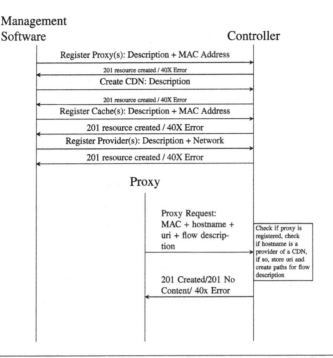

Figure 4.6 Northbound API interactions.

designed an additional private RESTful Northbound API, which is used only by these proxies. Before that, such proxies must be registered within the application through the public API exposed to OpenNaaS.

Once proxies and delivery networks have been set up through the public Northbound API as shown in Figure 4.6, the SDN application programs network devices to redirect HTTP requests targeted at a content provider towards the closest proxy. The proxy then uses the private Northbound API to notify the SDN application about the requested resource. If such a resource is not to be handled by any CDN, the controller gets in charge of routing the message to the origin server. Otherwise, the application programs a bidirectional flow from the proxy to the most appropriate cache that holds such a resource. If the resource is being requested for the first time, the application is responsible for choosing the most appropriate cache (according to the operator's policy) and programming the associated flow. Since this provokes a cache miss, the content must be downloaded from the origin server, but it will be available for future requests.

The validation of the system was carried out with DASH-encoded streams which not only allowed ensurance of the correct behavior of the system but also produced a side effect related to quality measurement since video playback results cannot be affected by the proposal.

4.4 Results

This section evaluates the architecture introduced in previous sections. Evaluation is split in two. First, the impact of IoT video encapsulated in DASH on CCN networks is evaluated. Next, our SDN-based solution for DASH video caching is shown. As already introduced, the results are related to the usage of DASH and the ICN adoption.

To obtain these results we have made use of the PlanetLab [62] orchestrated with NEPI [64] as well as the GaiaLab [63], part of the Enabling SDN ExperiMentAtion in WiReless Testbeds exploiting Future Internet Infrastructure in South KoRea and Europe (smart-FIRE) [71] project.

4.4.1 IoT Video Transport Impact on CCN

IoT networks are characterized by being constrained. As pointed out in Section 4.3.1, not only the data transmission is constrained but to a high extend also the processing power and the storage capabilities are scarce. These limitations shape the way video has to be processed and transmitted. If video is shaped to fit into IoT-constrained networks, it means that low bit rate is expected and what is more important, in the case of AVC, small Network Abstraction Layer Unit (NALUs) are expected. Those NALUs can be directly transported on top of DASH, that is, finally retrieved and transported on top of CCN. The authors in [72] introduce the evaluation of such a scenario. The performance obtained is unavoidably and predictably lower than the performance of typical host-to-host and IP-to-IP scenarios as a consequence of the small fragment imposed by the IoT nature.

Taking as a reference the experimentation architecture exposed in Figure 4.2 and thanks to its design, the power of in-network caching is shown, since the tree design with a root node avoids the interaction with the HTTP server after the in-network cache is filled. The different bit rates available are streamed to the leaf nodes (1, 3, and 6)/ causing "Cache Miss" on each of the network nodes in the CCN network. Consequent video retrievals produce a "Cache Hit" in the CCN network that reduces the retrieval time.

In addition, authors considered two different approaches to the video delivery. One non–time restricted, in which the media stream is completely recovered even if it produces pauses on the playing process. Finally, restricted time recovery was evaluated in which any media chunk that has not been retrieved on time is discarded, producing video bizarre effects and jumps in video stream. The former might correspond to VoD video streaming, while the latter might correspond to live content in which timing is essential.

Tables 4.1 through 4.3 represent the measurements obtained from streaming the three different pre-encoded bit rates of the video, i.e. 38, 58, and 79 kbps as in a VoD environment. From the tables it is concluded that the packet size restriction denotes itself as a definitive

Table 4.1   Experimentation with 38 kbps Bit Rate

BYTES	N1MISS	N1HIT	N3MISS	N3HIT	N6MISS
32	1686.0579	1665.3442	3407.6152	3392.2056	3676.5575
48	668.7148	658.0254	1547.9900	1555.4722	3268.7163
64	805.0028	752.0781	1325.8520	1421.4504	2530.6700
92	379.1102	359.1160	1092.7789	1099.6510	1723.3172
128	306.9185	278.9557	734.1414	798.5298	1137.7130
192	257.4231	232.9028	297.1269	324.3383	758.9249
256	276.7760	238.0714	372.0721	385.3181	573.7049
512	203.9962	168.3308	214.3915	228.7382	292.8262
1024	167.3246	162.0324	165.8882	162.3756	205.8480

Table 4.2   Experimentation with 58 kbps Bit Rate

BYTES	N1MISS	N1HIT	N3MISS	N3HIT	N6MISS
32	2118.1373	2097.1697	2877.0511	2902.7176	6209.2617
48	1628.9004	1594.7934	1031.5177	1029.8626	4323.3920
64	1081.4020	1055.4288	739.2264	774.2104	3873.9569
92	864.1175	839.6645	543.2547	547.4849	2841.0136
128	689.2486	619.7700	579.5825	661.6352	1699.3918
192	451.7747	381.5972	666.6065	703.0290	1293.6931
256	361.5066	309.5498	486.8731	480.0760	989.8218
512	247.8281	210.3649	204.0162	232.4632	459.0574
1024	160.1961	160.1581	214.0908	209.5092	329.0974

Table 4.3   Experimentation with 79 kbps Bit Rate

BYTES	N1MISS	N1HIT	N3MISS	N3HIT	N6MISS
32	3319.0994	2976.7420	4836.4743	FAILED	9085.2073
48	1310.2709	1265.1468	1905.6432	1946.8194	6186.4586
64	1645.4683	1636.9633	1339.5937	1614.4556	4829.9400
92	1141.9833	118.8758	7120.2064	729.1005	3085.2014
128	181.3390	159.6204	1607.7552	1590.2523	2898.6299
192	563.3972	491.3653	787.9775	845.9573	1540.1011
256	418.9757	381.1457	526.5776	529.0240	1392.3111
512	296.6672	256.7014	242,0550	278.1475	685.3320
1024	166,8538	159.7497	289.3108	277.0339	427.8385

factor on video retrieval performance with an exponential increase in retrieval time related to packet growth. There is also an expected relation between the number of CCN nodes in the path to the media source. Another factor not so easily identifiable that affects the performance is the DASH chunk size. The DASH chunk size was set to exactly one video frame; though video frame depends not only on resolution but also on the uniformity of the picture encoded. The more uniform the picture, the bigger the blocks taken for encoding. Anther typical influential mechanism in frame size is prediction, but as stated in Secton 1.4.1, that mechanism was deactivated on purpose. This effect can clearly be seen in Figure 4.7 from frame 400.

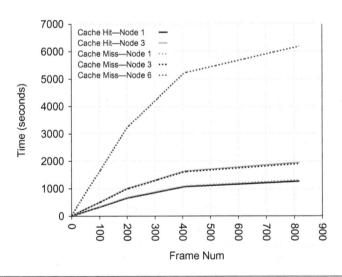

Figure 4.7   Results of 79 kbps transmission with 48 bytes limited network packet size.

| (a) Frame 602. | (b) Frame 603. | (c) Differences frame. |

Figure 4.8   Visual result frame 603 with a bit rate of 79 kbps.

The live performance is evaluated by testing different timeouts per chunk. If a chunk is delayed beyond its display time it is discarded and replaced with the last available chunk (something common in video decoders). The effect on the visuals is still images; depending on the kind of video content, the effect on the human eye is almost imperceptibl. But effects on PSNR are predictably bigger. Figure 4.8 shows visually what has just been explained.

The video retrieval has been experimented with chunk timeouts ranging from 1 up to 10 s. First a study of the timeout effect with the more restrictive video bit rate (38 kbps) was performed to select what the more representative timeout to be used is. In general 1 second is too low, which causes massive frame discarded, taking into account that the average RTT in PlanetLab is over 300 ms. Even 3 seconds of timeout is not enough time for packet sizes below 92 bytes. In general the retrieval is successful for timeout values over 5 seconds. Even with that timeout, packet sizes below 48 bytes are unable to be recovered. Table 4.4 shows the results of the video retrieval with each packet size between 32 (missing) and 1024 bytes. The table shows the

Table 4.4   PSNR 5 Second Timeout Study with 38 kbps Bit Rate

	CACHE MISS	# CHUNKS	CACHE HIT	# CHUNKS	δ PSNR	δ CHUNKS
0048	21.1791	283	28.5491	277	−7.3700	−6
0064	28.8081	407	28.8295	404	−0.0214	−3
0092	28.8001	407	28.8342	410	−0.0341	3
0128	28.8093	409	28.8342	410	−0.0249	1
0192	28.8050	472	28.8231	613	−0.0181	141
0256	28.7987	600	28.8229	618	−0.0242	18
0512	28.8057	468	28.8231	612	−0.0174	144
1024	28.8559	695	28.9062	818	−0.0503	123

timing and the number of chunks retrieved with empty and full in-network caches as well as the difference in PSNR between both cases (δ PSNR). A negative δ PSNR implies an enhancement in the video streamed for the cache hit.

After assessing the timeouts a deeper study is done with fixed 5 seconds timeout, limited to the packet sizes in the IoT range, meaning from 48 to 128 bytes. These experiments were repeated several times per size and the successful retrievals are shown in Tables 4.5 and 4.6. As can be seen, not only the Cache Hit case

Table 4.5 PSNR 5 Second Timeout Study with 58 kbps Bit Rate

	CACHE MISS	# CHUNKS	CACHE HIT	# CHUNKS	δ PSNR	δ CHUNKS
0048	30.0619	691	30.1837	803	−0.1218	112
0064	29.8381	359	29.8830	348	−0.0449	−11
0064	30.1323	742	30.1973	817	−0.0650	75
0064	30.1353	747	30.1981	818	−0.0628	71
0092	30.0070	400	29.9999	385	0.0071	−15
0092	30.0134	408	30.0328	406	−0.0194	−2
0092	30.0099	405	30.0325	405	−0.0226	0
0092	30.1741	792	30.1981	818	−0.0240	26
0128	30.1910	810	30.1981	818	−0.0071	8
0128	29.9986	428	30.0198	574	−0.0212	146
0128	30.0100	406	29.9090	359	0.1010	−47
0128	30.0087	473	30.0216	613	−0.0129	140

Table 4.6 PSNR 5 Second Timeout Study with 79 kbps Bit Rate

	CACHE MISS	# CHUNKS	CACHE HIT	# CHUNKS	δ PSNR	δ CHUNKS
0048	30.9516	654	31.0540	713	−0.1024	59
0048	30.5292	305	30.7512	330	−0.2220	25
0048	30.9659	667	31.0742	730	−0.1083	63
0064	30.4641	300	30.4765	302	−0.0124	2
0064	31.0744	723	31.1878	809	−0.1134	86
0064	30.4198	293	30.4334	293	−0.0136	0
0092	30.5083	306	30.9169	393	−0.4086	87
0092	30.3773	286	30.3529	280	0.0244	−6
0092	30.4487	300	30.4516	296	−0.0029	−4
0128	30.5010	324	30.9535	410	−0.4525	86
0128	30.9153	458	30.9192	611	−0.0039	153
0128	30.9234	407	30.9535	410	−0.0301	3

takes less time to finish but also there is a side effect on the video fidelity in relation to the original, therefore, achieving an enhancement in the user QoE.

The experimentation gives a good idea of what the three technologies—IoT, DASH, and CCN—can achieve in conjunction. The experiments demonstrated that IoT-constrained video can be encapsulated into DASH streams to leverage onto the World Wide Web (WWW) and HTTP–mature and optimized technologies. The inclusion of ICN by means of CCN has demonstrated the versatility of the technology for small packet sizes, which in turn has been an interaction of two FI technologies. The choice of PlanetLab as a testing facility introduces constraints that can be alleviated only in real IoT environments such as the RTT.

4.4.2 Leveraging Software-Defined Networking for Information-Centric Video Distribution

The main purpose of the above-mentioned methodology is to demonstrate that introducing SDN-based communications in CDN allowing the introduction of the ICN concept does not affect transmission. The study focuses on HTTP video transmission due to its relevance in computer networks communication in the near future [73].

In order to evaluate our proposal we deployed (see Figure 4.9) a two cache system over GAIA lab which is widespread over University of Murcia campuses and is part of the smartFIRE [71] federation. Smartfire3 and gaia-cache act as caches using Squid [74] 2.7 configured as a transparent http cache. gaia-proxy acts as proxy. Smartfire2 acts as client and omf-gaia-1 acts as router to leave the SDN. All the switches represented in the scenario are OpenVSwitch [75] 1.4.2 based. Our floodlight cdn app runs as controller in ATICA building where the HTTP server (plexp. inf.um.es) for the video is also hosted. For our experiments we use the well-known Big Buck Bunny [76] video in 480p (http://plexp. inf.um.es/bitdash/bunny/) and 1080p (http://plexp.inf.um.es/ bitdash/bunny1080p/) repackaged into one DASH stream with just one quality option to avoid client side quality switch decisions

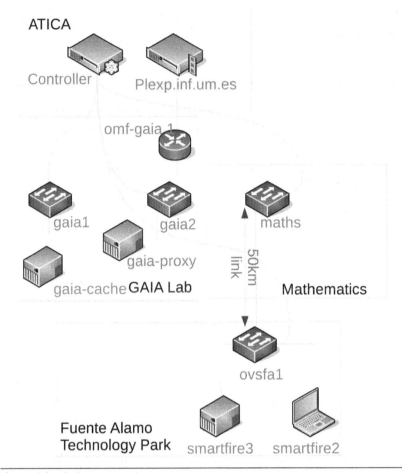

Figure 4.9 Evaluation scenario.

that would complicate the study. To perform the streaming we used the bitdash [77] javascript software.

The software deployed to evaluate this publication is publicly accessible:

- Floodlight cdn app: http://gitlab.atica.um.es/gn3plus/cdn-floodlight
- Proxy: http://gitlab.atica.um.es/gn3plus/gn3proxy

In addition a video of the whole solution in action can be seen at *http://tv.um.es/video?id=72371&idioma=es.*

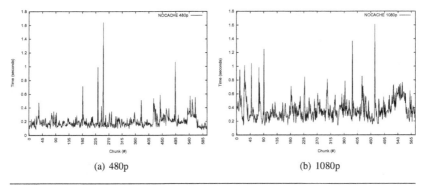

Figure 4.10   NOCACHE time spent by chunk.

First of all and as a basis for any conclusion, the NOCACHE case has been tested; in this case the client retrieved video directly from the server. Then, a typical caching system, NOCDN case, where clients use a predefined cache, was tested. Finally, this chapter's proposal, CDN case, was tested. The scenario is always the same with the same caching software but in the NOCACHE and NOCDN cases the SDN controller is programmed as an L2 learning switch.

Figure 4.10 shows the time spent for each video chunk, including MPD file retrieval as captured from the network device with a direct connection to the provider. As can be seen the retrieval of full hd video produces on average a 86% increase over the standard definition video.

In general, observing time spent per chunk in Figures 4.10 through 4.12 one can observe a plateau in the beginning and the end of the streaming process. The former is related to the need of downloading the initial mp4 that cannot profit from any frame prediction. The latter corresponds to approximate chunk number 495 corresponding to video time 8:15 where the credits start. Credits' lack of continuity in the image and thus intraframe prediction is complicated, increasing the produced bit rate, as can be clearly observed in Figure 4.13a. Figure 4.13 shows the bandwidth consumption by the video in SD (480p) and Full HD (1080p), including the HTTP and TCP headers.

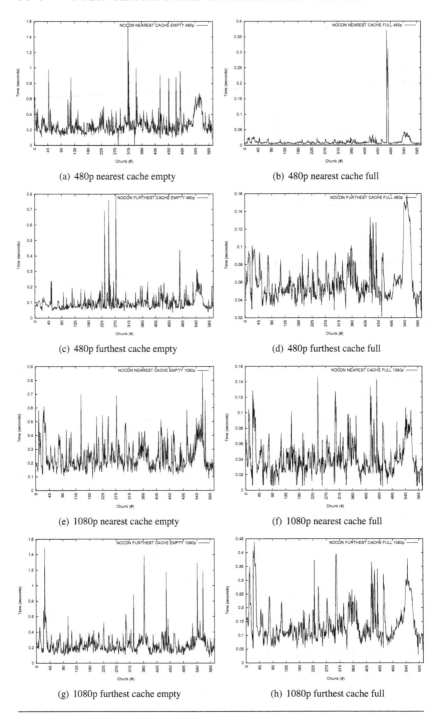

(a) 480p nearest cache empty

(b) 480p nearest cache full

(c) 480p furthest cache empty

(d) 480p furthest cache full

(e) 1080p nearest cache empty

(f) 1080p nearest cache full

(g) 1080p furthest cache empty

(h) 1080p furthest cache full

Figure 4.11   NOCDN time spent by chunk.

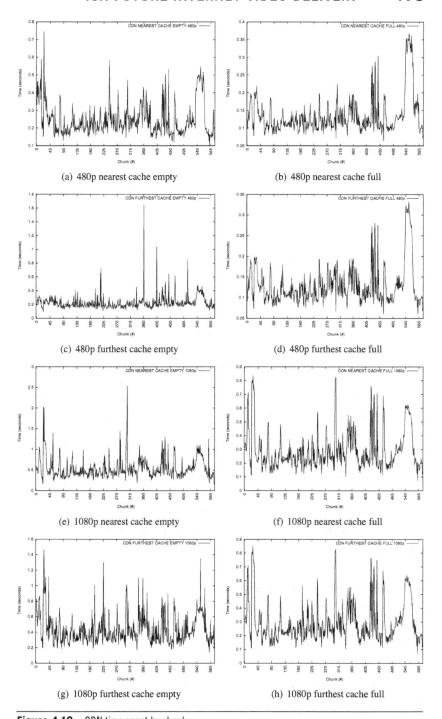

(a) 480p nearest cache empty

(b) 480p nearest cache full

(c) 480p furthest cache empty

(d) 480p furthest cache full

(e) 1080p nearest cache empty

(f) 1080p nearest cache full

(g) 1080p furthest cache empty

(h) 1080p furthest cache full

Figure 4.12 CDN time spent by chunk.

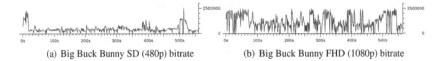

(a) Big Buck Bunny SD (480p) bitrate   (b) Big Buck Bunny FHD (1080p) bitrate

Figure 4.13   Source videos' bandwidth consumption.

Table 4.7   Streaming Client Experimentation Results

	SCENARIO	SIZE (BYTES)	DURATION (SEC)	DOWNRATE (bps)
	NOCACHE	239.891.091	557.0853	3.440.749.24
	NOCDN NEAREST CACHE EMPTY	272.818.063	557.4943	3.718.256.96
	NOCDN NEAREST CACHE FULL	251.868.515	556.9431	3.589.009.28
	NOCDN FURTHEST CACHE EMPTY	270.913.384	557.5884	3.712.513.85
	NOCDN FURTHEST CACHE FULL	267.028.208	556.8303	3.495.248.51
480p	CDN NEAREST CACHE EMPTY	260.976.310	557.0979	3.747.654.28
	CDN NEAREST CACHE FULL	261.228.081	557.0549	3.751.559.80
	CDN FURTHEST CACHE EMPTY	261.424.570	557.1328	3.753.856.57
	CDN FURTHEST CACHE FULL	261.486.751	557.4061	3.752.908.06
	NOCACHE	636.082.418	562.0779	9.049.156.84
	NOCDN NEAREST CACHE EMPTY	788.450.118	558.8509	10.749.786.39
	NOCDN NEAREST CACHE FULL	729.040.424	556.9834	10.236.077.81
	NOCDN FURTHEST CACHE EMPTY	778.085.776	556.2790	10.691.572.16
	NOCDN FURTHEST CACHE FULL	770.755.922	557.2209	10.641.239.69
1080p	CDN NEAREST CACHE EMPTY	758.935.705	558.8510	10.770.887.16
	CDN NEAREST CACHE FULL	757.339.029	557.5036	10.867.575.12
	CDN FURTHEST CACHE EMPTY	759.486.883	558.0119	10.798.963.94
	CDN FURTHEST CACHE FULL	759.542.011	557.4960	10.899.336.58

Looking at Table 4.7 one can observe that the NOCACHE cases take longer while downloading less data than the cases in which caches are involved. The reason for such a difference is that the caching systems involved force pipelining, allowing parallel chunk download. On the other hand, comparing the two caching system results, it is clear that there is no big negative impact in the streaming process when introducing the CDN approach and no negative optical influence was detected. Note that the CDN cases achieve higher download rates while the duration is slightly higher than the NOCDN cases due the initial delay produced by the proxy signalling through the REST API. SIZE takes into account from the TCP SYN message until the last TCP ACK, while the DOWNRATE takes into account only incoming packets.

Table 4.8 Per Chunk Experimentation Results

		AVERAGE	MIN	MAX	TIME X BYTE
	SCENARIO	SECONDS	SECONDS	SECONDS	SECONDS
	NOCACHE	0.197498	0.015380	1.640769	$4.743850 \cdot 10^{-7}$
	NOCDN NEAREST CACHE EMPTY	0.267126	0.010105	1.530754	$6.416318 \cdot 10^{-7}$
	NOCDN NEAREST CACHE FULL	0.012063	0.001702	0.370977	$0.289751 \cdot 10^{-7}$
	NOCDN FURTHEST CACHE EMPTY	0.099954	0.015448	0.799510	$2.400870 \cdot 10^{-7}$
480p	NOCDN FURTHEST CACHE FULL	0.060784	0.007177	0.158128	$1.460018 \cdot 10^{-7}$
	CDN NEAREST CACHE EMPTY	0.248177	0.084569	1.318234	$5.961166 \cdot 10^{-7}$
	CDN NEAREST CACHE FULL	0.135949	0.043314	0.367538	$3.265460 \cdot 10^{-7}$
	CDN FURTHEST CACHE EMPTY	0.228308	0.058377	1.641689	$5.483903 \cdot 10^{-7}$
	CDN FURTHEST CACHE FULL	0.129685	0.040498	0.330555	$3.115006 \cdot 10^{-7}$
	NOCACHE	0.367902	0.048682	1.611874	$8.836923 \cdot 10^{-7}$
	NOCDN NEAREST CACHE EMPTY	0.247191	0.011223	0.862217	$5.937470 \cdot 10^{-7}$
	NOCDN NEAREST CACHE FULL	0.040887	0.001584	0.146064	$0.982087 \cdot 10^{-7}$
	NOCDN FURTHEST CACHE EMPTY	0.243131	0.032143	1.488471	$5.839946 \cdot 10^{-7}$
1080p	NOCDN FURTHEST CACHE FULL	0.141554	0.006983	0.439054	$3.400088 \cdot 10^{-7}$
	CDN NEAREST CACHE EMPTY	0.540110	0.068427	2.534798	$12.973320 \cdot 10^{-7}$
	CDN NEAREST CACHE FULL	0.293359	0.041201	0.835532	$7.046423 \cdot 10^{-7}$
	CDN FURTHEST CACHE EMPTY	0.437545	0.139201	1.976934	$10.509730 \cdot 10^{-7}$
	CDN FURTHEST CACHE FULL	0.294563	0.039546	0.852044	$7.075341 \cdot 10^{-7}$

In order to have experimentation results not dependent on either web browser or streaming library, another set of experiments was performed in which each chunk (including the MPD file) was downloaded independently, not in a video streaming process. This approach also avoids the cloaking of any possible drawback of the solution caused by web browser caching or pipelining systems, meaning that the NOCACHE system is also using one TCP flow per chunk. Table 4.8 shows the statistics per chunk. As can be seen in the AVERAGE column, representing the average time in seconds for a chunk to be retrieved in each scenario, the CDN cases had an increase in time from 7% to 1026% but not exceeding in any case more than 0.2 second per chunk, an acceptable value when considering that the average size of a chunk is approximately 416 K for 480p and 1.2 M for 1080p, which is a 1 second video chunk. The best conclusion extracted is that for our CDN (0.41% to 4.608% enhancement in the nearest cache case) the distance between cache and client is not relevant when compared with the non-CDN case (80% to 246% enhancement in the nearest cache case). The reason for this result comes from the Northbound API signaling that becomes an important part of the consumed time per chunk. This results are for 1 second video chunks which means that fine grain video scrolling is possible without downloading extra data and reducing power consumption. Bigger video chunks would decrease the relevance of the Northbound API signaling while smaller video chunks would not have sense for human vision, although they would influence caching fragmentation.

The TIMExBYTE column adds the chunk size as input element. Caching systems offer the poorest results in cache miss (empty cache) cases, while best results are achieved with the same systems in cache hit (filled up cache) cases. There is, as expected, also a slight performance decrease when introducing this paper solution in front of the NOCDN caching, but there is still enhancement over the NOCACHE case. This measurement is considered to be relevant since chunk sizes differ due to Variable Bit-Rate (VBR) video coding and thus the effect in terms of TCP/HTTP overload differs as does the burden of northbound signaling.

An effect distinguishable in Figures 4.11 and 4.12 and corroborated by the values in Table 4.7 is that the time spent is in general higher with empty caches when accessing the nearest cache than when accessing the furthest cache. That is clearly related to the topology of the experiment (Figure 4.9) where the client's furthest cache is the provider nearest cache and vice versa and also nearer to the SDN controller. On the other hand, with full caches the behavior is more in line with what is expected observing reduced time for the nearest cache case.

The data plotted in Figures 4.10 through 4.12 corresponds to the summary of Table 4.8 which means that was recovered with the python script and not with the web-based video player. As can be observed, there are several impulses on the line, meaning that the retrieval of the chunk took more time than the average. We cannot find a regular pattern for these increases but looking at Figure 4.14 one can see that it is directly related to the emptiness of the caches so we can assume that the squid system is introducing the delay.

The experiments have validated the asumption that introducing SDN in order to be able to transparently provide ICN capabilities as a service is feasible and the penalty is negligible with results similar to what a transparent proxy achieves.

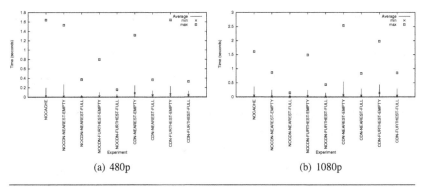

(a) 480p   (b) 1080p

Figure 4.14   Average, min, and max time per experiment.

4.5 Future Research

4.5.1 About VoD, Live Video Streaming, and Adaptive Streaming in the FI

The convenience of the all-HTTP in terms of simplicity and homo-geneity reduced the impact of specific streaming protocols which were kept only for certain cases as the multicast streaming. Nevertheless, the adoption of HTTP has a burden in terms of bandwidth consumption caused by the ancient TCP transmission, requiring at least double bandwidth as the bit rate of the stream. When Future Internet–based architectures come into play a new paradigm comes with them which implies in most of the cases breaking the already established standards. That case requires revisiting old techniques such as the use of containers with hint tracks but oriented to facilitate the naming of the content instead of the recent deployment-oriented alternatives.

When it comes to ICN, the applicability to VoD based on HTTP is quite clear and has been widely researched and analyzed. Adaptive streaming on the other hand is already an open field that generates some challenges mainly due to the in-network caching approach, while live streaming is in fact a special case of adaptive video streaming.

A common issue of ICN in general is the time a cached version of the content is considered valid and how the video content should be updated from the source. In environments like SDN, the controller can supply network statistics to the ICN networking nodes, simplifying the decision taking.

Another open issue for video streaming in general is the use of Digital Rights Management (DRM) and control over who can access the content and when. If the content is served from caches along the path it is not clear how a content provider can enforce its policies and charging mechanisms without the involvement of the network. For this kind of scenario, the use of NFV in conjunction with SDN for creating an overlay ICN network that is finally controlled by the content provider is, from our point of view, key for the adoption of ICN in production environments.

This kind of overlay also introduces other possibilities. In the case of live streaming, the ID/locator separation can be used to interconnect different ICN overlays belonging to multiple providers that would at last facilitate users' access to other providers' content, thanks to the local scope of the locators and the global scope of the IDs. The ID/locator needs also to be taken into account for the mobility scenarios of 5G where video is expected to be the main source of bandwidth consumption.

4.5.2 About Congestion and Caching Issues Related to FI

Caching mechanisms in ICN differ from the legacy CDN caching mainly because of the local scope of the ICN decisions. Each ICN node makes the decisions on the basis of the requests received and the data delivered. This approach simplifies the deployment of caches since any network entity is a potential caching node (in-network caching). What is more important, this system usually deploys the content near the client which potentially produces cache hits and avoids having a single point of failure [78]. Nevertheless, other issues are introduced such as how to implement caching hierarchies, how to reduce the number of copies of the same content on the caching system, and last but not least how to control who can access the information when the caching is done outside the provider network.

Multiple studies have been done [79] on the performance of and the best locations for CCN caches. Those studies are limited by the complexity of the calculations and are usually theoretical or applied to small deployments (two-level trees) [78,80]. Profiting from Future Internet architectures and in particular the flexibility of SDN, wider and more rich real deployment scenarios for CCN deployment could be used where different cache size as well as replacement and storage policies, among others, could be studied, also profiting from the centralized control to enhance cache deployment.

Several studies have addressed the size and performance of in-memory ICN caching. Something also to be taken into account with regards to ICN in-network caching when ICN becomes a NFV chain element is the waste of host memory while caching elements not of relevance for the actual ICN network usage. That waste of memory would avoid launching other NFV elements on the same host. Limiting the amount of information stored in-memory in the ICN routing elements would allow virtualization techniques such as memory ballooning to reduce the assigned memory and dynamically grow the size when really needed. We foresee this kind of study being necessary for the emergence of ICN on top of the Future Internet since both elements will probably be partly virtual.

Distributed caching is directly related to the concept of hierarchical routing, since to store a limited number of copies of each data element (at least one) implies that the routing to reach the element must be deterministic. In ICN in general two hierarchical concepts are applied, tree-based and Distributed Hash Tables (DHT). In hierarchical tree-based topologies each node has information on the elements in the same level of the architecture, its child nodes, and how to get to the parent. This approach reduces the amount of routing information needed by each node. In DHT and in Hierarchical Distributed Hash Tables (H-DHT) on the other hand, structures are in charge of sharing hash key information in a certain domain among the nodes belonging to that domain where all the nodes share the information, avoiding a single point of failure. In the case of H-DHT the parent domain has all the information on each of its subdomains.

Cooperative caching was dismissed in the last years of the 20th century for big populations. Authors stated [81] that "a single proxy cache can provide the same benefits." These results are still supported by the use of Facebook's caches [81]. In that sense, applying cooperative caching to ICN is expected to produce the same effect for big enough populations or organizations. On the other hand, and thanks to the centralization of the Control layer, these mechanisms which were complex to coordinate can be again inspected with the cooperation of the Network Controller.

The use of hierarchical naming structures is a well-known approach for ICN. The strings are concatenated similarly to how URI locators are created. With this approach any client is deterministically able to construct the URI of the desired resource which finally implies a certain path, reducing the burden of routing protocols. On the other hand, any modification to the data implies a modification of the URI which in turn could generate a high rate of URI discard.

The ICN architectures in general specify the routing within the domain and leverage an already established routing mechanism, such as BGP, for the inter-domain, while others develop their own inter-domain routing mechanisms [81]. Recently much effort has been put into migrating existing solutions to the new SDN paradigm. These efforts might be slightly redirected in order to increase the performance of ICN-based routing, increasing the network level taken into account from network to application.

The separation of identifier and locations for the Future Internet approach also introduces an open field for improvement in terms of inter-domain ICN routing. These systems, such as HIMALIS or MOFI, have addressed networking layer routing in a way similar to how local identifiers are used for local routing and prefixes are appended when the traffic must leave the domain. The ICN routing could profit from this approach to integrate and map the inter-domain mechanism and the content naming. Authentication mechanisms are among the issues pointed out by the community for the future of identifiers in Future Internet. It is clear that if an authentication mechanism for identifiers is achieved, the naming mechanism of ICN could profit from it by simplifying the identification of the content provider when the device moves location.

Recent proposals such as ID-Net [82] already intend to address the separation of ID and location and also make this ID independent from the networking mechanism by allowing some plug-ins through to the applications to directly employ the ID as the source and destination of the communication.

Since the appearance of HTTP-based video streaming mechanisms, several studies have tried to enhance or optimize the caching algorithms, taking into account that the content is video. In [83] the author defines a mechanism to adapt video bit rate while still trying to be fair with the caching mechanisms. In the case of [84] one must be aware that the caching mechanism is video being cached. So the former is client managed and the latter is cache managed. Along this same line, authors in [85] propose ICN awareness while caching

HTTP-based adaptive video content, and to that end, the authors define a DASH-INC cache that among other characteristics is able to modify the list of available providers in the MPD, introducing itself in the list if the content is already cached, the content is cached at the highest rate, and transcoding is done when lower bit rates are requested afterwards.

Thanks to the multiple descriptions and representations of DASH video streams, different storage algorithms can be designed to decide whether a video chunk needs to be stored in a certain cache or not. The decision could be made locally, as is usually done in ICN in-network caching, or taking into account neighbor information. Also algorithms based on the role that the ICN node has in the network can be envisioned. In that sense, a node within the core of the network could just store the initial chunks of each content to speed up the buffer fill time on the clients while the leaf nodes would try to store the whole media based on usage as usual. These kinds of approaches are even more interesting when Scalable Video Coding (SVC) comes onto the scene since it is clear that the base layer of the encoded video is needed for video decoding while the enhancement layers become irrelevant and worthless without the former. In such a case the core network elements would always try to store the base layer of the video which in turn is also the one with lower bit rate, thus storing a bigger number of unique contents.

4.5.3 ICN and Network Transport

Future Internet technologies in particular the split between ID and Locator facilitates mobility and the adoption of ICN which provides with a routing mechanism oriented to content clearly call to the usage of non-connection oriented transport or at least avoiding the Keep-Alive option used in TCP. Removing the Keep-Alive means introducing the TCP-Handshake in each transaction, therefore for small pieces of content makes sense to use datagrams, which taking into account the size of jumbograms in IPv6 (which wouldn't find any of the old objections to be widely and finally deployed in the FI) could be as big as 65 KB. This direct mapping between the network layer data unit and the application layer data unit would simplify the indexation of content by the ICN but also could lead to large object poisoning if small caches are in place [86].

Nevertheless, the link layer can trim this size to a more realistic 1500 octets. In such cases and for bigger chunk sizes aggregation on each ICN routing node is needed. On the other hand adopting datagram transmission reduce the complexity for the ICN, it is true that could lead to other side effects such as the need of retransmission. So it seems that connection oriented transmission is still desirable and TCP is the *de facto* standard nowadays but not the only alternative and other protocols like Stream Control Transmission Protocol (SCTP) could come into play.

Although it is clear that IPv6 is already deployed and available and not a feature of the FI, it is also true that it is not still the major protocol used at network level, indeed, most of the software running on top of the Internet is assumes IPv4. Backward support for IPv4 is mandatory and expected to continue there. In that sense the deployment of network adaptors in a transparent way is a need that will have to be taken into account in the near future, in particular, we foresee the inclusion of this adaptors as one of the functions to be chained in NFV on top of SDN and in coordination with the CCN nodes that will be deployed in the same way.

4.5.4 On How Policies and Regulations May Affect ICN

Copyrights and regulations relative to video authorship, such as the Digital Millenium Copyright Act (DMCA) [87], apply to video delivery in general and to Future Internet in particular. The usage of in-network caching already could be considered a copyright infringement but ignoring that, when a real infringement is found, reported, and substantiated, there must be a mechanism to remove all the incidences of that content in the network. So there are questions about how such a mechanism should be achieved while working with overlay networks and highly distributed network caching; again we envision the centralized control of SDN as the key feature to achieve this kind of control.

Another typical issue with video streaming is to ensure that the receiver of the content is a rightful recipient. The use of overlay networks with networking nodes and where caching is done for certain over-the-top (OTT) content and where the users can connect only after going through a successful authentication and authorization process leverages this kind of security. On the other hand multi-layer authentication and token-based authorization, such as OpenID Connect (OIDC), could be used to enforce authorization in ICN network nodes thanks to the previous network bootstrapping authentication.

4.6 Conclusions

As we have stated throughout the present chapter, video streaming operations on the Internet are quite simple operations from the end-to-end point of view but place complex requirements on intermediate elements of the network when trying to achieve certain levels of efficiency and effectiveness. As with general content transmission, the problem is stated as a simple producer and consumer scheme with intermediate caching functions. However, it is not trivial to instantiate such schemes in current networks and it is not easy to adapt current endpoints to a new operation mode.

We have delved into current architectures and proposals to evolve networks and overcome their limitations, with emphasis in the differentiation of content identification and specific underlying data transmission. We found that they do not meet the main requirements but can be used as a base to provide the necessary mechanism to achieve a basic, efficient, and effective video streaming service in the network. It is the case of ICN architectures, which are specifically designed to abstract the network from addresses and concentrate their operations into the content they are transmitting.

That said, our proposal follows the same principles as other ICN architectures to define simple and powerful content transmission mechanisms that are then instantiated onto the network by using SDN capabilities. Our proposal abstracts both producer and consumer endpoints from dealing with specific content-oriented operations by adding adapters to ingress and egress points of the network, so servers and clients can use the widely used HTTP protocol while the intermediate elements use a specific ICN solution adapted to SDN functions. Moreover, our solution exploits the benefits of adding caching elements to the network, so the content that is delivered through a certain path will stay available for some time in all intermediate elements of said path. This way, posterior requests will be resolved instantly by those nodes of the network without having to reach and contact the server that is initially streaming the content.

Providing everyday objects and any real-world object connectivity is accomplished by the IoT initiative. These devices have special characteristics and constraints that make their inclusion into networks a challenge. Both access networks and the Internet are suffering with the integration. At the same time, video has been and will probably continue being the main source of network traffic. Therefore in this chapter an evaluation of IoT capable video streams delivered on top of DASH have been evaluated, in a worldwide deployment on top of PlanetLab, employing a disruptive ICN approach such as the one proposed by CCN.

The results have shown how DASH and CCN fit perfectly together for the backbone network (such as the Internet), reducing the overhead produced by IoT network aware video traffic with small video chunks by encapsulating them all together. Also CCN has demonstrated to be useful by its message fragmentation capability to be transmitted on top of IoT networks. Finally, DASH is not a trend anymore but a reality and its simplicity in terms of content distribution is a key factor for the video-growing tendency.

Experimentation has shown a reduction in the performance in comparison with host-to-host IP-based communication. Nonetheless, the benefits offered by CCN while replacing IP for IoT environments are also visible. CCN flexibility in comparison with ossified IP approach and in-network caching capabilities facilitate the adoption to the constrained environment with the former and content delivery optimization with the later, as demonstrated with the experiments from previous sections.

As a matter of fact DASH is the key technology to make video transmission independent of the transmission mechanism and thus Future Internet friendly, no matter if we speak about ICN or IoT or—why not—embrace both.

In future work, we will continue with the evaluation of HIMALIS [4], as introduced above, and INP [88,89]. These architectures may also benefit to IoT general workloads and specific content and video delivery. Furthermore, we will redo the study presented here from the beginning with more streaming clients and bigger topologies. In addition, we plan to produce the same study with SVC in the line of [90] but deployed on top of PlanetLab. Thanks to SVC's scalability properties, better cache storage policies could be taken into account to enhance the cache hit ratio. As a consequence, we plan on applying SVC to our ICN SDN–based solution so that finally we can perform the experimentation with the full architecture.

Acknowledgments

This work is possible partially due to the Spanish National Projects PERSEIDES TIN2017-86885-R granted by the Ministry of Economy and Competitiveness of Spain (including European Regional Development Fund [ERDF] support), as well as the ANASTACIA Project that has received funding from the European Union's Horizon 2020 Research and Innovation Programme under Grant Agreement N° 731558.

Glossary

AHS   Adaptive HTTP Streaming. 9, 40

AVC   Advanced Video Coding. 8, 13, 22, 40

CCN   Content-Centric Networking. iii, v, 2, 6, 10, 11, 14–18, 22, 23, 26, 36, 38, 40

CDN   Content Delivery Network. 3, 7, 10, 11, 13, 18–22, 26, 35, 40

CDNaaS   Content Delivery Network as a Service. iii, 2, 10, 11, 19, 40

CIF   Common Intermediate Format. 13, 40

CoAP   Constrained Application Protocol. 10, 40

DASH   Dynamic Adaptive Streaming over HTTP. v, 9–11, 13, 14, 16–18, 22, 23, 26, 37, 39, 40

DHT   Distributed Hash Tables. 36, 40

DMCA   Digital Millenium Copyright Act. 38, 40

DNS   Domain Name System. 5, 6, 19, 40

DONA   Data-Oriented Network Architecture. 6, 40

DRM   Digital Right Management. 35, 40

EID   Endpoint Identifier. 4, 40

ERDF   European Regional Development Fund. 40

FI   Future Internet. iv, 2, 4, 9, 26, 35, 38, 40

FPA	Framework Partnership Agreement. 40
FPS	Frames per second. 13, 40
H-DHT	Hierarchical Distributed Hash Tables. 36, 40
HEVC	High Efficiency Video Coding. 8, 40
HIMALIS	Heterogeneity Inclusion and Mobility Adaptation through Locator ID Separation. 3, 5, 37, 40
HIP	Host Identity Protocol. 3–5, 40
HLS	HTTP Live Streaming. 9, 40
HTTP	HyperText Transfer Protocol. iii, v, 1, 7–11, 13–16, 18, 19, 21–23, 26, 28, 30, 35, 37, 39, 40
ICN	Information-Centric Networking. iv, 2, 4–6, 10, 11, 14, 18, 19, 21, 22, 26, 30, 35–40
IoT	Internet of Things. iii, 1–3, 5, 9–11, 14, 16, 22, 23, 26, 39, 40
IP	Internet Protocol. 3, 14, 18, 19, 22, 38, 40
IRTF	Internet Research Task Force. 5, 40
ISDN	Integrated Services Digital Network. 8, 13, 40
kbps	kilobit per second. v, vi, 11, 13, 23–27, 40
LISP	Location/ID Separation Protocol. 3–5, 40
MANO	Management and Orchestration. 20, 40
MOFI	Mobile-Oriented Future Internet. 3, 5, 37, 40
MPD	Media Presentation Description. 9, 13, 14, 16, 18, 28, 37, 40
MPEG	Moving Picture Experts Group. 8, 9, 40
NALU	Network Abstraction Layer Unit. 22, 40
NEPI	Network Experimentation Programming Interface. 14, 22, 40
NetInf	Network of Information. 3, 6, 40
NFV	Network Function Virtualization. 20, 35, 36, 38, 40
NICT	National Institute of Information and Communication Technology of Japan. 5, 40
NodeID	Node Identity Internetworking Architecture. 5, 40
OIDC	OpenID Connect. 39, 40
OTT	Over-the-top. 39, 40
PSIRP	Publish-Subscribe Internet Routing Paradigm. 6, 40
PSNR	Peak Signal Noise Ratio. vi, 13, 23, 25, 27, 40
PURSUIT	Publish Subscribe Internet Technology. 6, 40
QoE	Quality of Experience. 7, 26, 40

QP	Quantization Parameter. 13, 40
RANGI	Routing Architecture for the Next Generation Internet. 5, 40
RLOC	Routing Locator. 4, 40
RPL	Routing Protocol for Low Power and Lossy Networks. 10, 40
RTMP	Real-Time Messaging Protocol. 8, 40
RTP	Real-Time Transport Protocol. 7, 8, 40
RTSP	Real-Time Streaming Protocol. 8, 40
RTT	Round Trip Time. 16, 40
SAP	Session Announcement Protocol. 8, 40
SCTP	Stream Control Transmission Protocol. 38, 40
SDN	Software-Defined Networking. 2, 3, 10, 11, 18–22, 26, 28, 30, 35–40
SDP	Session Description Protocol. 8, 40
smartFIRE	Enabling SDN ExperiMentAtion in WiReless Testbeds exploiting Future Internet Infrastructure in South KoRea and Europe. 22, 26, 40
SVC	Scalable Video Coding. 37, 40
TCP	Transmission Control Protocol. 9, 18, 21, 28, 30, 35, 38, 40
URI	Uniform Resource Identifiers. 21, 36, 37, 40
VBR	Variable Bit Rate. 30, 40
VCEG	Video Coding Experts Group. 8, 40
VCR	Video Cassette Recorder. 7, 8, 40
VoD	Video on Demand. iv, 2, 23, 35, 40
WWW	World Wide Web. 26, 40
XML	eXtensible Markup Language. 18, 40

References

1. D. Meyer. The locator identifier separation protocol (lisp). *The Internet Protocol Journal*, 11(1):23–36, 2008.
2. D. Farinacci, V. Fuller, D. Meyer, and D Lewis. *Locator/ID separation protocol (LISP)*. Internet-draft, IETF, 2011.
3. R. Moskowitz and P. Nikander. Host Identity Protocol (HIP) Architecture, 2006. http://www.ietf.org/rfc/rfc4423.txt.
4. V. P. Kafle and M. Inoue. HIMALIS: Heterogeneity inclusion and mobility adaptation through locator id separation in new generation network. *IEICE Transactions on Communications*, E93-B(3):478–489, 2010.

5. H. Jung and S. Joo Koh. MOFI: Future Internet architecture with address-free hosts for mobile environments. *Telecommunications Review*, 21(2):343–358, 2011.

6. V. Jacobson, D. K. Smetters, J. D. Thornton, M. F. Plass, N. H. Briggs, and R. L. Braynard. Networking named content. In *Proceedings of the 5th International Conference on Emerging Networking Experiments and Technologies (CoNEXT'09)*, pp. 1–12, New York, ACM, 2009.

7. D. Trossen et al. Pursuing a Pub/Sub Internet (PURSUIT), 2011. https://cordis.europa.eu/project/rcn/95665/factsheet/en.

8. B. Ahlgren, M. DAmbrosio, C. Dannewitz, et al. Netinf evaluation. Technical Report FP7-ICT-2007-1-216041-4WARD/D-6.3, 2010. Deliverable D-6.3, 4WARD EU FP7 Project, http://www.4ward-project.eu.

9. M. Kang, J. Park, J. Shin, K. Nam, M. Shin, and J. Choi. Formal specifications for software-defined networking. In *Proceedings of the 7th International Conference on Future Internet Technologies, CFI'12*, p. 51, New York, ACM, 2012.

10. J. Ylitalo and P. Nikander. BLIND: A complete identity protection framework for end-points. *Lecture Notes in Computer Science*, 3957:163–176, 2006.

11. X. Xu. *Routing Architecture for the Next Generation Internet (RANGI)*. Internet-draft, IETF, 2009.

12. B. Ahlgren, J. Arkko, L. Eggert, and J. Rajahalme. A node identity internetworking architecture. In *Proceedings of the 25th IEEE International Conference on Computer Communications (INFOCOM 2006)*, pp. 1–6, Washington, DC, IEEE, 2006.

13. R. Atkinson, S. Bhatti, and S. Hailes. ILNP: Mobility, multihoming, localised addressing and security through naming. *Telecommunication Systems*, 42(3):273–291, 2009.

14. Information-Centric Networking Research Group. https://irtf.org/icnrg, 2012.

15. M. Brunner, H. Abramowicz, N. Niebert, and L. M. Correia. 4WARD: A European perspective towards the Future Internet. *IEICE Transactions on Communications*, E93-B(3):442–445, 2010.

16. T. Edwall et al. Scalable and adaptive Internet solutions (SAIL), 2011. http://www.sail-project.eu.

17. V. Dimitrov and V. Koptchev. PSIRP project—publish-subscribe Internet routing paradigm: New ideas for Future Internet. In *Proceedings of the 11th International Conference on Computer Systems and Technologies and Workshop for PhD Students in Computing on International Conference on Computer Systems and Technologies*, pp. 167–171, New York, ACM, 2010.

18. T. Koponen, M. Chawla, B.-G. Chun, A. Ermolinskiy, K. Hyun Kim, S. Shenker, and I. Stoica. A data-oriented (and beyond) network architecture. *SIGCOMM Computer Communication Review*, 37(4):181–192, 2007.

19. P. Martinez-Julia and A. F. Gomez-Skarmeta. Using identities to achieve enhanced privacy in future content delivery networks. *Computers and Electrical Engineering*, 38(2):346–355, 2012.
20. A. Austerberry. *The Technology of Video & Audio Streaming*. Burlington, MA, Elsevier, second edition, 2005.
21. B. Bing. *Next-Generation Video Coding and Streaming*. Hoboken, NJ, Wiley, 2015.
22. C. Perkins. *RTP. Audio and Video for the Internet*. Boston, MA, Addison-Wesley, 2003.
23. H. Schulzrinne, S. Casner, R. Frederick, and V. Jacobson. RTP: A transport protocol for real-time applications. *Technical Report, Internet Engineering Task Force*, July 2003. Standard, RFC 3550.
24. M. Handley, C. Perkins, and E. Whelan. *Session Announcement Protocol*. RFC 2974 IETF, 2000.
25. M. Handley, V. Jacobson, and C. Perkins. SDP: Session description rotocol. Technical report, Internet Engineering Task Force, July 2006. Proposed Standard, RFC 4566.
26. H. Schulzrinne, A. Rao, and R. Lanphier. Real time streaming protocol (RTSP). *Technical report, Internet Engineering Task Force*, April 1998. Proposed Standard, RFC 2326.
27. H. Parmar and M. Thornburgh. Adobe's real time messaging protocol. *Copyright Adobe Systems Incorporated*, pp. 1–52, 2012.
28. ISO/IEC 11172-2:1993—Coding of moving pictures and associated audio for digital storage media at up to about 1,5 Mbit/s—Part 2: Video. Technical report, 1993.
29. ISO/IEC 13818-2:2013—Generic coding of moving pictures and associated audio information—Part 2: Video. Technical report, 2013.
30. ISO/IEC 14496-2:2004—Coding of audio-visual objects—Part 2: Visual. Technical report, 2004.
31. Theora Specification. Technical report, 2004.
32. ISO/IEC 23008-2:2013—High efficiency coding and media delivery in heterogeneous environments—Part 2: High efficiency video coding. Technical report, 2013.
33. A. Grange and H. Alvestrand. A vp9 bitstream overview. 2013.
34. H. Schwarz, D. Marpe, and T. Wiegand. Overview of the scalable video coding extension of the H.264/AVC standard. *IEEE Transactions on Circuits and Systems for Video Technology*, 17(9):1103–1107, 2007.
35. Y. Ye and P. Andrivon. The scalable extensions of HEVC for ultra-high-definition video delivery, *Multimedia IEEE*, 21(32):58–64, 2014.
36. Y. Chen, Y.-K. Wang, K. Ugur, M. M. Hannuksela, J. Lainema, and M. Gabbouj. The emerging MVC standard for 3D video services. *EURASIP Journal on Advances in Signal Processing*, 2009(1):786015, 2009.
37. A. Biernacki and K. Tutschku. Performance of HTTP video streaming under different network conditions. *Multimedia Tools and Applications*, 72(2):1143–1166, 2014.

38. H. Rani and E. Bansal. A review on HTTP streaming strategies in media streaming. *International Journal of Engineering and Computer Science*, 4(8):14033–14035, 2015.

39. Roger Pantos. Http live streaming—draft-pantos-http-live-streaming-19. Internet-draft, 2015. draft-pantos-http-live-streaming-19.

40. ISO/IEC 23009-1:2014—Dynamic adaptive streaming over HTTP (DASH)—Part 1: Media presentation description and segment formats. Technical report, 2014.

41. Christian Timmerer and Christopher Mueller. HTTP streaming of MPEG media. In *Proceedings of the Multimedia Communication 26 (MM 26)*, pp. 1–4, 2010.

42. M. G. Michalos, S. P. Kessanidis, and S. L. Nalmpantis. Dynamic adaptive streaming over http. *Journal of Engineering Science and Technology Review*, 5(2):30–34, 2012.

43. D. Singer, editor. *ISO/IEC 14496-12:2005 Part 12: ISO Base Media File Format*. Geneva, Switzerland, International Organization for Standardization, 2005.

44. IEEE 802.16 Working Group et al. IEEE standard for local and metropolitan area networks—part 16: Air interface for fixed broadband wireless access systems. *IEEE Std*, 802:16–2004, 2004.

45. G. Montenegro, N. Kushalnagar, and D. Culler. Rfc 4944. transmission of ipv6 packets over IEEE 802.15. 4 networks. online. September 2007.

46. J. Hui and P. Thubert. Rfc 6282. Compression format for ipv6 datagrams over IEEE 802.15. 4-based networks. 2011.

47. J. Granjal, E. Monteiro, and J. Sa Silva. Security for the Internet of Things: A survey of existing protocols and open research issues. *IEEE Communications Surveys and Tutorials*, 17(3):1294–1312, 2015.

48. T. Winter, P. Thuber, B. Brandt, et al. Rfc 6550: Ipv6 routing protocol for low-power and lossy networks. *Internet Engineering Task Force (IETF) Request For Comments*, 2008.

49. C. Bormann, A. P. Castellani, and Z. Shelby. COAP: An application protocol for billions of tiny Internet nodes. *IEEE Internet Computing*, 16(2):62, 2012.

50. B. Astuto. A survey of softwareDefined networking: Past, present, and future of programmable networks. *IEEE Communications Surveys & Tutorials*, 16(3):1617–1634, 2014.

51. H. Kim and N. Feamster. Improving network management with software defined networking. *IEEE Communications Magazine*, 51(2):114–119, 2013.

52. G. Mulligan. The 6lowpan architecture. In *Proceedings of the 4th Workshop on Embedded Networked Sensors*, pp. 78–82. New York, ACM, 2007.

53. T. Watteyne, F. Chraim, N. Sarmicanic, C. Jian, and K. S. J. Pister. Video transmission over a standards-based wireless multi-hop sensor network. *E-LETTER*.

54. E. Callaway, P. Gorday, L. Hester, J. A. Gutierrez, M. Naeve, B. Heile, and V. Bahl. Home networking with IEEE 802.15. 4: A developing standard for low-rate wireless personal area networks. *Communications Magazine, IEEE*, 40(8):70–77, 2002.

55. Itu-t recommendation, h264 advanced video coding for generic audiovisual services. *ISO/IEC*, 14496, 2007.

56. T. Wiegand, G. J. Sullivan, G. Bjontegaard, and A. Luthra. Overview of the h. 264/avc video coding standard. *IEEE Transactions on Circuits and Systems for Video Technology*, 13(7):560–576, 2003.

57. J. Ortiz Murillo, M. Ransburg, E. Martínez Graciá, M. Sablatschan, A. F. Gómez Skarmeta, and H. Hellwagner. Towards user-driven adaptation of h. 264/svc streams. 2010.

58. GPAC Multimedia Open Source Project. https://gpac.iob/.

59. C. Müller and C. Timmerer. A vlc media player plugin enabling dynamic adaptive streaming over HTTP. In *Proceedings of the 19th ACM International Conference on Multimedia*, pp. 723–726, New York, ACM, 2011.

60. J. Reichel, H. Schwarz, and M. Wien. Joint scalable video model jsvm 0. *Joint Video Team of ITU-T VCEG and ISO/IEC MPEG, Doc. JVT N*, 21, 2005.

61. PYCCN NDN Bindings for Python. https://github.com/named-data/pyccn.

62. B. Chun, D. Culler, T. Roscoe, A. Bavier, L. Peterson, M. Wawrzoniak, and M. Bowman. PlanetLab: An overlay testbed for broad-coverage services. *SIGCOMM Computer Communications Review*, 33(3):3–12, 2003.

63. P. Martinez-Julia, A. J. Jara, and A. F. Skarmeta. Gaia extended research infrastructure: Sensing, connecting, and processing the real world. In *Proceedings of the TridentCom 2012*, pp. 3–4. Springer, 2012.

64. A. Quereilhac, M. Lacage, C. Freire, T. Turletti, and W. Dabbous. NEPI: An integration framework for network experimentation. In *Proceedings of the 19th International Conference on Software, Telecommunications and Computer Networks (SoftCOM)*, pp. 1–5, Washington, DC, IEEE, 2011.

65. B. A. A. Nunes, M. Mendonca, X.-N. Nguyen, K. Obraczka, and T. Turletti. A survey of software-defined networking: Past, present, and future of programmable networks. *Communications Surveys & Tutorials, IEEE*, 16(3):1617–1634, 2014.

66. B. Heller. OpenFlow switch specification. *Current*, 0:1–36, 2009.

67. Pitt, Dan. https://www.opennetworking.org/?p=2332&option=com_wordpress&Itemid=316.

68. L. Peterson, S. Baker, and M. De Leenheer. XOS: An extensible cloud operating system. In *Proceedings of the 2nd...*, 2015.

69. Floodlight project. http://www.projectfloodlight.org/floodlight/.

70. The OpenNaaS project. http://opennaas.org.

71. SmartFIRE Enabling SDN Experimentation in Wireless Testbeds exploiting Future Internet Infrastructures in South Korea and Europe. http://eukorea-fire.eu/.

72. P. Martinez-Julia, E. T. Garcia, J. O. Murillo, and A. F. Skarmeta. Evaluating video streaming in network architectures for the Internet of things. *Innovative Mobile and Internet Services in Ubiquitous Computing (IMIS), 2013 Seventh International Conference on*, pp. 411–415. IEEE, 2013.

73. Cisco. Cisco global cloud index: Forecast and methodology, 2014–2019. *White Paper*, pp. 1–41, 2014.

74. Squid: Optimising Web Delivery. http://www.squid-cache.org/.

75. Multilayer Open Virtual Switch Open vSwitch Production Quality. http://openvswitch.org/.

76. Big Buck Bunny. https://peach.blender.org/.

77. bitdash MPEG-DASH Players High Streaming Quality. https://www.bitmovin.com/bitdash-mpeg-dash-player/.

78. G. M. Brito, P. Braconnot Velloso, and I. M. Moraes. *Information-Centric Networks: A New Paradigm for the Internet*. Hoboken, NJ, ISTE Ltd/John Wiley & Sons, 2013.

79. H. Li and H. Nakazato. Optimal cache allocation for content centric networking, pp. 1–18, 2014.

80. Y. Wang, Z. Li, G. Tyson, S. Uhlig, G. Xie, Y. Wang, Z. Li, G. Tyson, S. Uhlig, and G. Xie. Design and evaluation of the optimal cache allocation for content centric networking. *Design and Evaluation of the Optimal Cache Allocation for ContentCentric Networking*. 65(1):95–107, 2016.

81. A. Ghodsi, S. Shenker, T. Koponen, A. Singla, B. Raghavan, and J. Wilcox. Information-centric networking: Seeing the forest for the trees. In *Proceedings of the 10th ACM Workshop on Hot Topics in Networks – HotNets'11*, New York, ACM, pp. 1–6, 2011.

82. H. Jung, W.-S. Lim, J. Hong, C. Hur, J. Lee, T. You, J. Eun, B. Kwak, J. Kim, and H. S. Jeon. IDNet: Beyond All-IP network. *ETRI Journal*, 37(5):833–844, 2015.

83. H. Yunfeng. Cache-friendly rate adaptation for Dynamic Adaptive Streaming over HTTP (DASH). PhD thesis, 2015.

84. D. H. Lee, C. Dovrolis, and A. C. Begen. Caching in HTTP adaptive streaming: Friend or foe? In *Proceedings of Network and Operating System Support on Digital Audio and Video Workshop*, pp. 31:31–31:36, 2013.

85. R. Grandl, K. Su, and C. Westphal. On the interaction of adaptive video streaming with content-centric networking. *2013 20th International Packet Video Workshop*, San Jose, CA, *PV 2013*, 2013.

86. Y. Thomas, G. Xylomenos, C. Tsilopoulos, and G. C. Polyzos. Object-oriented packet caching for ICN. In *Proceedings of the 2nd International Conference on Information-Centric Networking – ICN'15*, San Francisco, CA, pp. 89–98, 2015.

87. DMCA The Digital Millenium Copyright Act of 1998. https://www.copyright.gov/legislation/dmca.pdf.

88. P. Martinez-Julia and A. F. Gomez-Skarmeta. A novel identity-based network architecture for next generation Internet. *Journal of Universal Computer Science*, 18(12):1643–1661, 2012.

89. P. Martinez-Julia and A. F. Skarmeta. A lightweight and identity-based network architecture for the Internet of things. In *Proceedings of the Sixth International Conference on Innovative Mobile and Internet Services in Ubiquitous Computing (IMIS) 2012*, pp. 711–716, Washington, DC, IEEE, 2012.
90. J. Lee, J. Hwang, N. Choi, C. Yoo, and B. Labs. SVC-based adaptive video streaming over content-centric networking. *KSII Transactions on Internet & Information Systems*, 7(10):2430–2447, 2013.

5

STORAGE-CENTRIC NETWORKING

M. BALA KRISHNA AND DANIEL CORUJO

Contents

5.1 Storage-Centric Networking

5.1.1 Introduction

Storage management and storage-centric network (SCN) support data accessibility and storage for large-scale users in the network. In sensor networks, the storage and retrieval operations are based on node queries of the local and external storage systems. The cluster head nodes (CHs) and base stations (BS) coordinate with local storage systems. The controlling entities enable the data flow rate between the local sensor groups and external storage systems. The data archiving systems apply intelligence and lossless local data aggregation techniques at the CHs to store the aggregated data and support limited user queries. Data retrievals from the hash-based data-centric storage systems [1] with pre-determined routing mechanisms consume high energy in large-scale multi-hop networks. Storage and retrieval latency [2] is reduced by using the temporary data storage nodes, forwarding and mobile nodes moving towards the destination. The storage capacity, abstraction, restoration and data dissemination are improved by using techniques such as cache proxies, lazy-offloading and multi-resolution, server virtualization and cloud-based systems [3–8]. Solar-powered battery systems enhance the storage capacity of sensor nodes. The services of message ferries and nodes with flash memories implement the storage-centric approach in sensor networks to enhance the data collection process in intermittently connected regions. This technique significantly sustains the environmental hazards in large-scale sensor networks. Intelligent storage systems support autonomous data storage and retrieval operations [9] in virtual and cloud-based storage systems. Encrypted storage [5] systems prevent unauthorized data access and retrievals by malicious nodes. SCN addresses issues related to secure data storage, retrieval management and analysis in large-scale dynamic networks. Storage-centric networking [1,3,4,6,8,10] enables the authentic nodes consisting of high-capacity, flash memory and distributed storage systems to provide the secure storage and retrieval operations in the network. SCN enhances the data abstraction and distributed storage capabilities of wireless sensor networks, mobile ad hoc networks, Internet of Things and the edge computing systems.

This chapter describes various storage-centric networking and storage management techniques based on structural, operational, virtualization, cloud computing, edge computing and the security features of wireless sensor networks, cognitive radio, IoT and cloud-based environments. Section 5.1.2 explains the classification of storage-centric networking systems. Section 5.1.3 explains the characteristic features of storage-centric networking. Section 5.1.4 explains the storage and sub-storage systems in terms of the storage attributes, system models and storage data hierarchy. Section 5.1.5 highlights the functionality of intelligent storage and scheduling systems. Section 5.1.6 explains storage virtualization using dynamic mapping and software-defined storage systems. Section 5.1.7 explains the security features in various storage systems. Section 5.1.8 explains the WPAN and WBAN storage systems. Section 5.1.9 highlights the significance of storage systems in mobile and wireless networks. Finally, Section 5.1.10 concludes the chapter and gives direction for future extension.

5.1.2 Classification of Storage-Centric Networking Systems

With increasing data demands from large-scale networks, the storage reusability and computational capacities of database servers are also enhanced. The storage-centric systems are classified based on

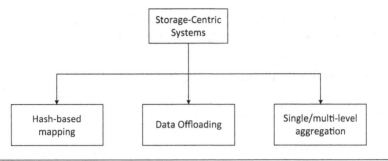

Figure 5.1 Classification of storage-centric systems.

(i) hash-based mapping [1,8,11], (ii) data offloading [3,12,13], (iii) single/multi-level aggregation [1,5,14,15], (iv) distributed data dissemination [6,16,17], (v) storage virtualization [7,9,18,19], (vi) cloud storage systems and [14,20–24] and (vii) programming, intelligence-based systems [9,11,21].

Figure 5.1 illustrates the classification of storage-centric systems. The graph embedded sensor labeling approach [1] with a virtual polar coordinate system consists of *n–hop* neighborhood storage nodes with dynamically re-constructible routing tables and shortest route paths. This approach reduces the complexity of data sensing and aggregation in SCN. Distributed hash tables are updated to (i) map the newly added data storage nodes, (ii) improve the reusability of memory blocks and (iii) manage the spanned pages to sustain the frequent node failures. This technique reduces the data speed and storage performance of the system. NAND/NOR-based flash memory storage systems [3] with MicaZ platforms profoundly improve data speeds, enabling the nodes and server to perform the simpler storage abstractions. Platform-independent and tunable memory systems reduce the storage capacity exhaustion rates by intelligently archiving the old data into the external storage centers at BS. Data abstractions performed at the sensor storage increase the computational complexity and operational costs. In low-powered sensor systems, the data abstractions with long file names and directory lists are carried out at the BSs to offload the sensor nodes. Lazy-offloading [4] and cross-partition data offloading techniques by data mules minimize the load of sensor nodes. Lazy-offloading persists the data flow in the neighborhood of sensor nodes until the storage and energy levels are maintained within the threshold limit. The size of data chunks is varied with slot

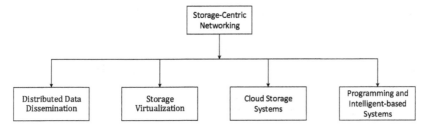

Figure 5.2   Classification of storage-centric networking.

availability at local storage units while the nodes remain in the active states such that the remaining energy is more than the estimated data storage lifetime. The cross-partition data offloading by data mules enables the use of free storage slots advertised by active sensor nodes, and the storage gradients trace the unused memory slots. Thus, the data mules redistribute the data chunks towards the available storage slots of the system. Figure 5.2 illustrates the classification of storage-centric networking.

Asynchronous data storage systems [5] support fine granular fingerprints of the sensing regions in terms of node updates and energy consumption rates in distributed databases. The aggregator nodes at storage systems apply the persistent additive homomorphic encryption scheme and verify the minimum dominating neighborhood set of communicating nodes to prevent the rate of eavesdropping in the network. A storage-centric system [6] with multi-data resolution technique considers the lightweight trees with low transmission power in which the route paths share the adaptable transmissions (idle schedules) of the descendant nodes and minimize the energy consumption rates in the network. Data-centric storage systems [8] apply the coding techniques to partition the network into sensing and storage zones. The collector and manager nodes coordinate with each other to build an efficient storage system. Deployment of solar-powered software and hardware embedded sensors in a large-scale sensor network [10] achieves minimum energy and optimal coverage in remote regions. The storage system consists of data collection, file access, replication and scattering layers to collect the data using the basic, reliable and balance modes of the network. The data replication layer is activated or deactivated based on the number of active sensor nodes per layer and the node energy levels maintained within the threshold limits. Stream-oriented storage-centric systems [25] with a rechargeable battery and application-specific storage policies reduce node failure rates. The storage functionality and energy tradeoffs are improved by decoupling the data storage and processing methods.

5.1.3 Characteristics of Storage-Centric Networking

The main features of storage-centric networking are given as follows:

1. Enhance the data storage and abstraction capabilities of the sensor nodes and storage entities in heterogeneous, disruptive, distributed and dynamic networking environments.
2. Storage-centric systems are scalable, intelligent, secure and tangible to deal with the node social behavior.
3. Enable large-scale sensing systems to sustain the link failures, random access and operational delays during the read or write operations.
4. Framework with access controls and operational logs for managing the workloads between multiple vendors and heterogeneous users. Handle data aging, redundancy, error handling and improve data accessing with fast retrieval rates.
5. Create the virtual storage system that separates the logical and physical storage to minimize the operational cost.
6. Data loss arises due to the lack of coordination between the sensing, networking and data storage components.

5.1.4 Storage and Sub-Storage Systems

Storage-centric network management is comprised of (i) cache and disk management, (ii) content-based management, (iii) load management, (iv) compressive sensing and (v) energy-efficient network resource management. Secure storage systems support authentic data access from the distributed data centers in the network.

5.1.4.1 Attributes of Storage and Sub-Storage Systems   Specifications of storage and sub-storage systems [12,16,26–30] include an efficient storage design plan, well-defined policies for storage partition and management. The data sharing between virtual and real cache centers at local and global storage affects system performance. Storage efficiency is a function of data preservation and replica management schemes in intermittent connected networks. The primary characteristics of distributed storage and sub-storage systems are given as follows:

- *Data streams*: Data streams are the collection of data items [11] based on the time of creation, modification (by source), number of updates, payload size and value.
- *Data staleness*: Date staleness represents the suspended data items and time duration that the message is not repeated in the current session. Data staleness is the number of non-updated message contents and differentiated key values of input and output data items.
- *Data reputation rate*: The degree of semantic replications of user queries matched at storage centers in the request/response system during the peak traffic hours is called the data reputation rate. The forwarding nodes store the data replicas for the time period specified by the local servers. The node, cluster, BS or centralized database server measures the data reputation rate. Local replication indicates the number of data classes matched at the cluster or BS, and global replication indicates the number of data classes matched at the data storage centers.
- *Reliability*: Storage reliability depends on pre-defined levels of information sharing between the data resources. Security, access control and user restriction policies applied to real and virtual database servers ensure data reliability in large-scale storage systems.
- *Scalability*: Scalability deals with the extension of data centers and efficient management of distributed data centers (real and virtual) for data retrievals and user queries in the system.
- *Mapping function*: The mapping function indicates the linear function and relevant parameters to represent the data and key fields in database tables. Mapping functions such as content hashes with the longest path keys [28] determine the frequency of identical contents and their stamps to balance the load of real and virtual data storage centers.
- *Aggregation rules*: The data aggregation levels depend on user density and network size. The step-wise aggregation rules (applied at the node, CH and BS) construct the data

aggregation tree. Data aggregation rules are based on data partitions allocated to the application by the centralized storage server. The matched data classes are aggregated separately with different replacement policies as per the data resources, such as HTTP that uses regular ports for streaming data, social media and marketing applications. This simplifies the process of data retracing from the source node. Aggregation rules for shared data implement indiscriminate replacement policies suitable for autonomous applications.

- *Average failure rate*: The number of rolled back transactions by the application servers and the queries with unmatched contents recorded at local and global data storage servers over a time period indicate the failure rate.
- *Operational costs*: Operational costs depend on the storage model, power consumption levels and management policies of the storage system. The communication cost is based on the number of queries, node size and latency in the network. Operational cost is a function of successful packet delivery rate and allocated bandwidth. The power control mechanisms [30] introduce redundancy of encoded data packets that are proportional to the number of active nodes in the system. The nodes broadcast posterior probability distributions to ensure the existence of appropriate active packets at the cache centers to compensate for the lost data.
- *Workload management policy*: Workload management addresses the issues of user query bandwidth, delivery latency and throughput rates at multiple local and global database servers. The policies restrict the number of read and write access rights allocated per user. The pre-defined priority workload policy [27] is applicable for small-scale systems with limited capacity, whereas the weighted fair workload policy supports the replacement of old data packets with new arriving requests, and the number of storage slots shared by the application.

5.1.4.2 Storage and Sub-Storage System Models The four-layer storage model is comprised of a data collection layer, a gateway and synchronization layer, data mapping, an analysis and storage management layer and a service provider management layer. Figure 5.3 illustrates the architecture of storage and its relevant sub-system model in SCN.

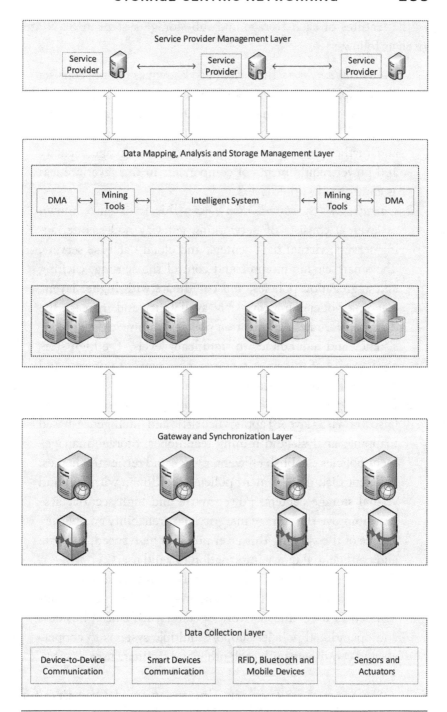

Figure 5.3 Architecture of storage and its relevant sub-system model in SCN.

The features of each storage and sub-storage system in SCN are given as follows:

1. *Data collection layer*: This layer (also known as layer 1) consists of a comprehensive network of heterogeneous sensing devices such as sensors, actuators, RFIDs, cameras and mobile devices. The components initiate periodic and dynamic events, and further generate voluminous data. The storage capacity and power requirements of components in this layer are less as compared to other layers.

2. *Gateway and synchronization layer*: This layer (also known as layer 2) consists of networking gateway nodes, interfacing servers, virtual data centers, and cloud database servers. The smart clouds monitor and control the sensing, caching and aggregation policies, and coordinate with higher layers. Caching policies improve data availability and increase the data transfer rate. Aggregation policies minimize data inconsistency and redundancy to significant levels. The high storage capacity of this layer requires more power to process and manage the voluminous data.

3. *Data mapping, analysis and storage management layer*: This layer (also known as layer 3) applies heuristic and intelligence-based mapping, analysis and mining techniques. Storage management policies establish efficient storage and retrieval schemes. Efficient disk management policies coordinate with real and virtual storage centers. Power-aware and high security systems improve the storage integrity, data reliability and performance of the system. This layer provides high speed, accurate and confidential query information to valid users.

4. *Service provider management layer*: This layer (also known as layer 4) coordinates with the vendor-specific business models mapped to users and network operators. Enhanced business and operational policies support multiple systems to cooperate and coordinate with each other and facilitate user services.

5.1.4.3 Sub-Storage Data Hierarchy The system model consists of a multi-layered data storage system. The query-based data storage system (layer 1) collects the raw data using online or offline queries from authentic users (valid IDs registered with the application server) and cache buffers. The application-server storage system (layer 2) stores the user query data at the cache and process the query at application servers. The servers classify the data and store them at application-specific backend servers. Data mining and operational storage systems (layer 3) apply heuristic, intelligent and soft computing techniques to derive knowledge and information from the data. This is further transferred to the centralized server storage system with vendor-specific storage systems. A centralized storage server defines the service list and incentives, and assigns them to large-scale users interacting with lower layers of the system.

5.1.4.4 Content-Based Storage in Sensor Network   Data storage [16,26] in a large-scale sensor system supports efficient storage planning, minimizes the data replica rate and ensures data availability at the server. Directed acyclic graph-based data storage and management systems apply map reducing technique [26] to arrange the data using popularity and arrival rates. The content-centric storage partitioning scheme [27] considers the hierarchical naming structures to improve the scalability. The information sharing levels with storage transparency are maintained at the local cache enters. The missing information is retrieved from the forward information and storage backups at the cache centers. The new user request is forwarded with an entry in the pending interest table to track the data flow. This technique provides an immediate response to user queries with similar interests.

The dynamic distributed ring-based storage model [28] preserves the namespace locality and applies the hash-based partition in the network overlay for fast data retrieval. The histogram-based load balancing scheme dynamically maps the namespace at metadata servers and reduces the data lookup rates. The lookup messages are assigned keys to accommodate the longest path for new files and directories in the sub-tree of metadata servers. The keys use content hash fields to reduce the number of data replicas with identical contents received from the distinct paths. The load balancing is achieved by mapping the contents to virtual storage centers and evaluating the local range density histograms at metadata servers. Data preservation consumes energy in large systems and requires efficient flow controlling algorithms [29] that push and retrieve the data quickly. The distributed data storage model applies erasure coding technique [30] and randomized power control schemes to sustain random node failures and reduce data redundancy. As per sensing history, long-term data storage [12] requires efficient placement of data replicas. Data accessibility rates are increased in the balanced mode using the round robin technique. This approach maximizes the availability of old sensing events with the new arriving data in the system.

An IoT-based data storage system [14] efficiently combines the sensing aspects of sensors and RFID devices. This system applies object entity mapping to store the data of IoT objects. Greedy-based distributed data storage systems [31] support the resilient capacity of active nodes to retain the data replicas using the memory tables of neighborhood nodes. Software-defined networking storage systems [17] apply the hierarchical cluster approach to ensure dynamic resource allocation and support multiple tenants since the physical and data storage systems are decoupled from each other.

5.1.4.5 Storage-Centric Systems This section highlights the effects of (i) power awareness of distributed data storage centers, (ii) block level allocation policies, (iii) load management and (iv) compressive sensing techniques of the storage-centric systems. Figure 5.4 illustrates the power-aware, load management and sensing in storage-centric systems.

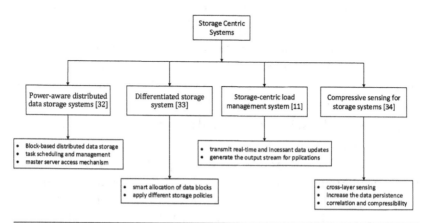

Figure 5.4 Power-aware, load management and sensing in storage-centric systems.

5.1.4.5.1 Thermal Power–Aware Hadoop-Based Storage Model High-speed internet applications with distributed data centers consume more power due to the data replica retained at storage centers, online task scheduling, and disk rotational rate. Hence, the thermal-aware storage and scheduling systems minimize the heat recirculation rates. Thermal power–aware Hadoop-based storage [32] system considers the following features:

1. A block-based distributed data storage system supports large files that are equally partitioned into uniform blocks and stored at different data centers. This reduces the storage complexity and fastens the file accessing mechanisms. The data centers coordinate with nodes and minimize deadlines of data replications.

2. A minimum cost task scheduling and management system supports file accessing queries related to large files. The queries are scheduled into multiple smaller requests to retain the multiple data blocks of user requests and maintain the temperature of the cooling system within the threshold level. This reduces the storage and accessing complexity. The thermal model ensures the hot air from data nodes is circulated back to the air conditioner to maintain the temperature levels at the inlet and outlet of the system.

3. The master server controls the access mechanism of renamed files at data nodes existing in multiple racks of the storage center. The data and their replicas are separately stored at different storage racks in different machines. Since the data nodes with varying bandwidths process the user requests, the data blocks from the same storage racks minimize the latency.

4. Cost and edge connectivity affect the recirculating temperatures of data nodes processing the data replicas with the same deadlines. The task performing cost of data node, source and sink nodes are proportional to the node distances from the air conditioning and the data storage systems.

The power-aware performance coefficient depends on the ratio of total power consumed by the active data nodes and the air conditioner system. The data replicas at nodes with highest outlet temperatures are moved to the idle data nodes with identical replicas and deadlines. The data replicas with different deadlines ensure that the cumulative performance coefficient does not exceed the maximum temperature constraint.

5.1.4.5.2 Differentiated Storage System   The performance of the storage system depends on (i) smart allocation of data blocks supported by the semantic information and (ii) different storage policies for I/O classes related to file systems, directories, table indexes and database logs. The distributed data storage system [33] addresses the block-level classifiers for featuring the data classification and management policies. Smaller bandwidth allocation for low latency user priority, accessing the small files and high bandwidth allocation for large files are primarily considered in this scheme. The storage classifiers indicate the data integrity, encryption and replication levels required by the file systems. The three-layered storage model consists of block layer at the bottom, file system layer in the middle and application layer at the top. The storage model with stateless classifier does not preserve the classes of each block and allows the de-duplication of data blocks from the same storage pool. The file classification system allows priority-based caching and eviction of data blocks based on disk seek rate and the user accessing requests (random and sequential). The data accessibility from non-volatile memory systems is addressed by the size-based selective allocation and eviction priority policies.

5.1.4.5.3 Upstream Storage-Centric Load Management System   High-speed data stream processing engines used in financial services transmit real-time and incessant data updates for online users and consider load balancing techniques to minimize data staleness. The storage-centric systems [11] support key sensitive queries of distributed database servers and address issues related to high-speed data streams. This further uses the semantic update to correct the data staleness in user queries. The update stream consists of data symbols and keys that indicate the current query status. In this model, the input stream periodically pulls the continuous queries to processing engines and generates the output streams for applications. The updates in a tuple-based stream provide the most recent modifications of input and output key values, and the window-based stream updates apply the relevant aggregate functions in relevant data fields that are grouped in the present window.

5.1.4.5.4 Compressive Sensing and Storage System   Distributed data storage [34] with compressive signal samples supports the cross-layer sensing and communication between the nodes. The signal sampling and signal recovery components evaluate the probabilistic broadcasting of wireless channels. Further, the sensor readings are rebroadcasted to avoid retransmissions by the source nodes. The data collector node retains the network readings by decoding the samples from the minimal data set. This model aims to increase the data persistence with sustenance to correlation and compressibility in large-scale WSNs. The data dissemination rate is improved with the minimum number of transmissions and decisions that are independent of the routing table size. The random sampling projections are defined using $K << N$ sparse signals, and the optimal signal recovery is achieved with the linear system of sparse signals. Node communication depends on the Euclidian distance and maximum transmission range with limited forwarding probability p. The probability giant value ensures that the maximum number of nodes have received the network broadcast messages. The N-dimensional random projection row matrix allows equal node probability to act as the source nodes and broadcast the messages to neighborhood nodes. The resultant M-independent row sparse matrix consists of a higher number of broadcasting source nodes stored as non-zero uniform entries existing in each row. This model facilitates the distributed data storage at the source nodes with the minimum number of broadcast messages. Table 5.1 indicates the functional aspects of mapping functions used in storage systems.

5.1.5 Intelligent Storage and Scheduling Systems

Intelligent storage systems [9] define the block-based storage interfaces to differentiate the storage services and stabilize the data transport between the node and storage systems. Cloud-based storage systems with intelligent matching and data sharing at multiple locations prevent the data loss. The layered cloud storage model coordinates with the network and storage infrastructure that is comprised of physical

Table 5.1 Mapping Techniques Used in Storage Systems

MAPPING RULE	ADVANTAGE	DISADVANTAGE
GEM (Graph EMbedding for sensor networks) [1]	Uses the data-centric approach and virtual polar coordinate to ensure the data accessibility of sensor nodes	Nodes preserve the states of dynamically changing neighbors
Symmetric and asymmetric mapping system [7]	Uses different mapping schemes for encryption, and aggregation schemes provide short-term and long-term storage for data replication	Complexities increase with the heterogeneity of the sensing devices
Task node time mapping [11]	Low-cost task rescheduling method maximizes the node performance	Rescheduling complexities increase in large-scale networks
Lineage and update mapping schemes [13]	Support the time-based data output streams and optimize the functionality of non–strictly correct query systems	Complexity increases for frequently revised queries in dynamic overloading systems
Constructive algorithm mapping [19]	Evaluates the matching of resource requests	Cumbersome for nodes with low storage levels and dynamic priorities
Object entity mapping [21]	Supports the foreign key constraints in cross-platforms	Complexity increases for large-scale sensing with multiple sensing units and RFID tags
Hierarchical bin packing [22]	User-specific service level agreements enable the load balance module and dynamically provide server instances and storage pool connections	Reduces the complexity of migration size in large-scale dynamic clusters of mobile and vehicular network traffic
Random virtual–physical machine mapping [29]	On-demand and live migration of virtual machines	Energy consumption increases in complex and dynamic overloading systems
Object storage device–based volume mapping [30]	Storage area is efficiently mapped as per user requirements	Data retrieval speed is not enhanced
Bucketing scheme for value-tag mapping [36]	Reduces information leakage and ensures user data privacy through encoded numbers supported by external storage nodes and servers	Compromised storage nodes and servers use false encoded numbers, and modify the false alarm rates and upper bound limits

and logical functions. The load balance is achieved by the collaboration of storage and metadata with distributed data domains and logical entities of the system. The storage service interface enables the access rights to authentic clients, and the storage convergence module applies the evolutional techniques to simplify the migration of storage objects across multiple users. Ant colony optimization with intelligent rules eases the data access and reduces the service overload at storage servers. Intelligent buffer management [33] applies the flag classification based on index types, file access patterns and I/O requests. The data buffer levels are varied to improve the processing speed.

5.1.5.1 Functionality of Intelligent Storage Systems The functional features of intelligent storage systems [18,21,33,35,36] are given as follows:

- Reduce the computational complexities of data access for the solid state disks and the hard disk drives
- Improve the data read/write speeds
- Sustain the high level I/O functionality and buffer management in terms of data reliability
- Allow sequential write operations on hard disk drives and minimize the random writes on solid state disks; this approach supports content locality and strong regularity of data blocks
- Apply lightweight application–dependent proxy re-signature schemes to preserve the data integrity of source nodes

The coupling [35] of solid state and hard disk drives efficiently improves the storage performance, overcoming shortages of solid state devices with NAND-gate flash memory and computational complexities of I/O operation on hard disk drives. Intelligent coupling balances the number of current read reference data blocks, and the frequently read blocks of I/O system. The system performance is analyzed using the similarity detection and delta variation levels of online read and write operations.

5.1.5.2 Smart Storage Systems The public auditing storage system [18] creates the proxy re-signature revocation scheme. Based on user integrity and source node performance levels, the data access permissions are reassigned. The reliable storage allocation system [21] derives the generating function in the upper bound, and the lower bounds are evaluated to allocate the redundant blocks in the storage system. Data optimization in cloud storage systems [36] applies the ad hoc approach to estimate the geometric space using attributes such as latency, operational cost and false tolerance level.

5.1.6 Storage Virtualization

In storage virtualization systems, the logical abstracts are directly linked to heterogeneous users with zero disruptions enabling free data flow between multiple hosts across the network platforms. Storage virtualization at the host considers the logical file abstraction from physical storage (files extracted from virtual disks) associated with the network. Storage virtualization [7,22] is based on resource utilization between the hosts and application servers. Storage caching [13] with hypervisor-based memory devices reduces the complexity of shared cache devices and performance isolation in virtual machines. The run-time application environment, guest operating systems and multiple storage systems are continuously analyzed in storage management. The effective cache ratio parameter identifies the actual cache space utilized by the virtual machines, and the fair share of each virtual machine is evaluated to yield the high accuracy rates.

5.1.6.1 Dynamic Mapping Techniques   Storage virtualization [37] shares the infrastructure with varying attributes such as access rights, throughput and latency rate. A storage virtualization system supports the sequential and random data accessing schemes along with historical data access. Thus, the dynamic mapping systems achieve efficient file management and dynamic storage with authorized access rights, steady inflow and outflow data streams in the cloud-based environment.

5.1.6.2 Software-Defined Storage Systems   Software-defined storage systems (SDS) [19] reduce storage and retrieval complexities in cloud computing applications. Storage virtualization in SDS aims to achieve (i) automation of shared storage systems with efficient data accessing policies and (ii) resource optimization by using object storage systems to sustain the large data growth. An object storage system with proxy and storage nodes allows the concurrent data access of multiple tenants with varying data filters. The advantage of data filters is that they allow the tenants to use only a fraction of a large data set to complete the task. This approach considerably increases the data flow and reduces the download traffic of tenants. The multi-stage and ready-to-use dynamic data accessing policies allow the flexibility to read or write the user codes based on object requests. The data policies in terms of domain-specific language are given as follows:

- State the options for DO and WHEN clauses of data caching.
- Set the data compression for large data files exceeding the limits.
- Modify the bandwidths for current I/O operations.

During the caching process, the user invokes the compiler. The SDS controller verifies the fulfillment of the WHEN clause of user policies and resets the caching data by invoking the relevant filters. The user data filter policies are verified and certified by the metadata and code management system connected to the application servers. The user requests are classified based on filters applied to reset HTTP headers that are further executed by the proxy storage nodes. The message-oriented middleware coordinates with the system and the service metrics are obtained from the data plane satisfying the WHEN clause that allows the disseminated data into the metadata stores.

5.1.7 Security in Storage-Centric Networking

Data security is the pre-requisite for storage-centric systems that ensure the integrity, privacy and confidentiality of user information. The storage capability of sensor systems is enhanced by the solar-powered battery to sustain the encryption methods and provide reliable data. The persistent encryption-based data storage model [5] elects the aggregator node with high processing and storage ability to implement the relevant encryption technique. The sensing period is reduced to prevent unauthorized access by adversary nodes. A distributed data storage system [38] improves the robustness of the sensor network and increases the integrity by partitioning the user data using erasure coding and a secret key sharing approach. Table 5.2 highlights the security and privacy features of existing data storage models.

Table 5.2   Security and Privacy Aspects in Storage Systems

SECURITY MODEL	FEATURES	LIMITATIONS
Tiny Persistent Encrypted Data Storage (TinyPEDS) [5]	Supports end-to-end security using symmetric additive homomorphism for data aggregation, and asymmetric additive homomorphism for long-term storage	Data dependence and integrity issues are not adequately addressed
Distributed data storage security model [39]	Supports fine-grained data distribution by defining the relevant access controls and privileges for personalized access to private data	Needs frequent updates to address safety issues in emergency situations
Dependable secure data storage scheme supporting dynamic integrity [38]	Ensures data integrity by generating the secret sharing of erasable codes	Complexity increases for group-wise communication and unreliable users in large-scale cloud-based systems
Verifiable privacy preserving bucketing scheme [40]	Prevents the disclosure of raw data storage and allows data encryption by the sensors. The pre-loaded keys are replaced with new one-way key and the hash function for the forwarding node	Works for queries within the specific range
Selective redundancy secret sharing model [23]	Supports security for large files using the retrievable intermediate variable file chunks based on optimal erasure codes, and reduces the operational cost	Becomes complex for distributed and autonomous systems
Secure and scalable mass storage model [15]	Multi-coefficient polynomial and internal padding scheme for secret sharing in distributed data storage system applies the Gauss–Jordan Elimination to define the coefficient matrix for accurate data retrieval	Complexity increases with context-aware services and data redundancy at multiple servers
Cloud storage auditing model [41]	Supports anonymity of cloud network users through verifiable public coding and public auditing by a third party to verify the intruders	Implements heavy keys and does not support data dynamics related to random modification, insertion and data deletion blocks
Group–based symmetric balanced incomplete block design [24]	Group–based data sharing scheme with common secure key suitable for public cloud systems that support anonymity of users	Does not provide a mechanism to identify single user meddling as multiple user identities

5.1.7.1 Security and Privacy in Storage Systems The security and privacy attributes considered in secure storage systems [5] are given as follows:

1. *Monitoring the range and behavior of attackers*: The attackers are active for a short duration and monitor the limited range. Hence, the secure storage system needs to monitor the behavior of attackers.
2. *Regional-based and device-specific data replication*: Due to variations in active- sleep periods and node battery levels, secure storage systems vary data replication rates.
3. *Adaptable redundancy factor:* Dynamically varies the data as per the average energy of sensing region and ensures that the redundancy levels are kept at the minimum.
4. *Storage energy cost*: The secure storage system restricts the data replications per region and minimizes the storage energy cost.
5. *Attacker sustenance range*: The minimum possible distance of active nodes from the attacker to forward the secure data.

5.1.7.2 Dependable Secure Data Storage System Large-scale storage systems with semantic computing and collaboration manage to retain the structure of I/O requests. Data partitions with lightweight security, erasure coding with algebraic signatures and secret key sharing methods [38] improve the robustness, confidentiality and integrity of distributed data storage systems. The hybrid data share generation, fast integrity and spot checking schemes analyze the false positive and negative indexes to identify the modified data stored in data center. The data shares are extended over multiple data blocks, and the false index analysis indicates the level of data falsification by attackers in the system. Threats caused by signature capture and eavesdropping of adversary nodes are prevented by the multi-party integrity checking conducted by the signature sharing information.

5.1.7.3 Secure Data-Centric Storage Systems Secure data-centric storage systems [23,40,42] allow the legitimate nodes to retrieve the data using resource type, access, and energy levels. A privacy-preserving storage framework [40] evaluates the user query range by encoding the data. A multi-storage IoT network [15] applies the secret sharing and internal padding techniques to improve the security and scalability. A secure data-centric cloud storage system [43] applies the authentication and encryption techniques to secure the data. This results in efficient data aggregation at the network level that enables the distributed data nodes to join or leave the session. A hybrid computational distributed data storage system [44] improves the data reliability and reduces the data recovery cost. A secure cloud storage system [41] enables the third-party service provider to verify the networking codes assigned to anonymous users and update the data. The proxy re-encryption scheme applies the same plaintext to define multiple keys and separate the real and proxy systems. An IoT-based storage model implements the stochastically time-based mathematical model [45] to evaluate the hypergraphs and map the links across the nodes in the network.

5.1.8 Storage in Personal Area Networks and Wireless Body Area Networks

Personal area network (PAN, IEEE 802.15.4) devices are used in WLAN (IEEE 802.11) for buffering the data packets in static sensor networks [4]. PAN devices act as delay-tolerant storage nodes for transferring the data to sink nodes. Trusted mobile nodes known as data mules use the internet links to bridge the transmission gaps between the source and sink node by connecting the IEEE 802.15.4 serial ports. PAN devices with low power and computational abilities do not sustain high storage and security protocol complexity. Data storage and accessibility in WPANs and WBANs require flexible policies with low storage and computational overheads. The secure data storage system [20] verifies the data and header information given by encapsulating the security payload of the underlying IP security system.

5.1.8.1 Secure Cloud-Based 6LoWPAN Storage Systems

6LoWPAN adaptive with ZigBee (IEEE 802.15.4) extends the services of smart IoT devices by reducing the header length. Secure storage systems [20] for IoT devices support IP and 6LoWPAN to ensure trusted communication between end-to-end devices. Thus, the data packets satisfy IPsec/ESP format and allow direct data transfer to IP stack. The cloud-based approach reduces the additional overheads and enables the storage of compressed authentication header (AH) information for successive data packets. The direct transfer of stored data packets in the ESP format avoids complex cryptographic processing. The UDP (user datagram protocol) header is attached to the secured ESP to store the data. Stored secure node association specifies the IP flows, and the compressed information of current and next headers are authenticated with the integrity check value. The initialization vector indicates the current security compression and the validation status of UDP/IP data packet uses the selected encryption scheme.

5.1.8.2 Storage Management in Wireless Body Area Networks The distributed data storage system for wireless body area networks (WBANs) [39] is comprised of sensitive and personalized information. WBANs include wearable and implantable body sensors, health monitoring sensors and RFIDs. WBAN systems are equipped with high security, privacy and location updates of active devices. Tiny sensors continuously monitor the vital information and send the updates to local servers for data aggregation, processing and data storage in cloud and application database servers. Further, the fine-grained information is transmitted to the centralized database server. The distributed systems operate in two modes: (i) the authentic remote users directly access the data and (ii) wearable sensors directly address authentic user queries. This approach requires fine-grained access privileges assigned to authorized persons. The security and access control lists depend on the (i) access residue and data shares received from the storage nodes, (ii) signatures received by the nodes, (iii) roles and privilege mappings allocated to nodes, (iv) cipher text policies and (v) the encrypted keys assigned to nodes. Storage confidentiality ensures that the distributed and differentiated access rights are assigned to valid authorities to access the sensitive data. The flexible access policies ensure low computational and storage overheads. Further, the security protocol complexity in WBAN depends on the coverage area and policies of external infrastructure. Thus, high security mechanisms with adequate activation steps are considered in the coverage area. In emergency and disaster services, verifiable access policies are considered. Dynamic data integrity using random verification checks of information leakage is sent to alert the users.

Fault-tolerant storage nodes and local servers facilitate the reliable data to end users. Data integrity in distributed storage systems [46] ensures that the user does not directly modify the data in the storage center, and the data dependability scheme sustains the data loss during intruder access. Secure data storage is initiated by the network administrator that considers the preloaded hash functions that are proportional to the number of body sensors attached to the patient. The trusted key center generates the group keys for selected authority such as specialist and healthcare staff attending the patient. The shareholders using the group key are validated by periodic broadcast challenging messages, and the replies received from users represent the data integrity of sensor nodes.

5.1.8.3 Cross-Layer Distributed Storage Systems   Cross-layer storage systems [47] use the bi-directional communication of metadata and apply the top-down and bottom-up storage approaches. In the top-down approach, applications estimate the file access patterns, distributed file locality and expected lifetime to optimize the data placement in storage layers. In the bottom-up approach, storage systems reveal the data location indexes and accessing schedules to ease the data navigation. The three-level storage system consists of workflow and runtime engines to coordinate with task schedulers and assign the application tasks to nodes. These nodes perform tasks such as file abstraction, file sharing, aggregation and file storage in the backend system. The bottlenecks of I/O file transfer are enhanced by the batch scheduler that balances resource allocation and defines the access patterns for file sharing. The POSIX storage interface constraints are reduced as follows: (i) allocate to single large application program interface (API); (ii) extend the support of additional ad-hoc API to improve the I/O speed; (iii) vary the selection of I/O routines; (iv) classify and rearrange the API block size with respect to metadata, journals, large file systems and small file systems; (v) apply the enhanced caching and precoding techniques and (iv) use the programming frameworks, replication management and file access control lists.

POSIX-compatible storage systems [47] for cross-layer data flow apply the incremental file system approach that coexists with flexible middleware and virtualized storage systems. The workflow storage model with intermediate file optimization and location-aware scheduling schemes enhances the cross-layer data flows. Storage benefits of big data management are estimated at the application level using the data sharing and navigation levels of the nodes. The workflow scheduler and storage program interface coordinate with each other. The metadata manager utilizes the file access patterns given by the workflow scheduler, and the performance of the storage access interface is enhanced by using the online or offline status of storage nodes. The storage nodes and the metadata manager maintain the retrieval and garbage collection of data blocks. For the new file, the block allocation is linearly incremented, and the unused space is collected using the asynchronous garbage collection method. Data storage is configured by the dispatcher node that supports the access patterns, replica management and space allocation to accommodate the new write operations into the data files. The storage access interface manages the file-related operations and cache management between the multiple data blocks. POSIX passes the storage hints (in the form of key value pairs) to application programs, and triggers the relevant modules and metadata manager for information retrieval accordingly.

5.1.9 Storage-Centric Networking in Mobile and Wireless Networks

The flash memory systems [3] in wireless networks apply intelligent techniques to reduce the complexities of data storage and data abstraction. These memory systems are flexible, simple, tunable, cost-effective and platform-independent.

5.1.9.1 Storage in Mobile and Wireless Networks Due to random mobility and non-uniform node distribution in a particular geographical region, data delivery and storage in mobile networks are challenged by intermittent connections between the source and server nodes. The message ferrying technique [2] utilizes the services of specially deployed mobile nodes to enhance data delivery in disrupted and sparsely connected regions of the network. The node-initiated and mobile ferry–initiated delivery modes are selected based on the crisis, topography of sensing zone and application service cost. The functional modes are explained as follows:

- *Node-initiated data delivery mode*: Proactive periodic updates are received from the adjacent mobile nodes connected to the message ferry mobile nodes. Here, the mobile nodes are allowed to move along the pre-defined route paths of the network.

- *Mobile ferry data–initiated delivery mode*: The mobile ferry nodes proactively reach the sparsely connected regions to receive the data from mobile sensor nodes. The message ferry nodes act as relay nodes and support the long-range communication.

The message generation rate and message drop rates indicate the data delivery levels in the network, and the work time parameter estimates the node task completion time. The message drop rates reduce the buffer outflow rates and the number of timed out message at ferry nodes. Mobile nodes send the service requests based on the communication distance between message ferry nodes while moving towards or away from the sensing zone.

5.1.9.2 Distributed Storage in Mobile Wireless Networks   The distributed storage in a mobile wireless network depends on the communication cost and cache content overheads in the system. The storage model [48] supports the node scenarios describing the (i) random arrival or leaving, (ii) incoming nodes with cached contents and (iii) repair of the lost cached contents when the storage nodes randomly leave the BS. The periodic use of erasure codes for content caching at the device storage level addresses the issue of content repairing. In a mobile wireless network, the random join and leave nodes with multiple coded symbols consider the following rules:

- The coded symbols satisfy the predefined expected arrival rates.
- The number of data storage nodes is always less than the number of mobile nodes associated with the BS.
- The requested file can be downloaded if the file retrieval is completed within the time duration for which the active storage nodes present in the cell are above the minimum required limit h.

Content repair is required when the mobile device frequently leaves the BS while storing the data contents. Hence, the rate of content repair in device-to-device (D2D) and BS depends on the number of storage nodes within the coverage area. If the number of storage nodes is high, then the content repair is performed at the D2D level using the data contents available with the current active mobile devices; otherwise, the BS performs the content repair. The repair cost is defined in terms of bandwidth, contact interval and the interaction probability between the communicating nodes and BS. The downloading cost per bit from the BS and adjacent nodes estimates the battery load in corresponding mobile devices. The storage classes with predefined nodes eliminate the risk of unused symbols used by the mobile nodes.

5.1.9.3 Storage Model in Cache-Aided Wireless Networks   The cache-aided wireless network [49] addresses the issues of storage and latency in wireless networks. The storage model determines the contents to be cached and efficiently delivers the requested contents in the storage capacity of edge nodes. The channel state information (CSI) from edge nodes, normalized delivery time, high signal-to-noise ratio (SNR) and degrees-of-freedom (DoF) define the delivery latency in non-interfering links. The edge transmission depends on the following policies:

- Efficient file mapping that satisfies the cache capacity constraints and allows arbitrary coding
- Applies fixed caching policy without inter-file coding
- Edge transmission policy depends on the demand vector representing the file requests mapped between the edge nodes and users
- Normalized delivery time determines the tradeoffs between the maximum delivery latency and cache capability of edge nodes
- Decoding policy depends on receiver demands, channel output parameter and the channel realization

The fractional cache size depends on the file storage used by the edge nodes and the total size of popular library files distributed over the shared wireless channel. The fixed caching policy for multiple transmissions is defined by the function that maps the user file to cache storage of the edge node. The cache-aided policy determines the channel outputs based on distributed entropy, interference alignment, and zero-forcing beamforming scenarios.

5.1.10 Conclusion and Future Directions

This chapter elucidates the characteristics of storage-centric networking and sub-storage systems. Distributed architectures, virtualization and security aspects in storage-centric systems are explained in detail. Intelligence-based techniques in storage-centric networking enhance the performance of data retrieval rates, quality of information, reliability and relevance of user demands. Storage virtualization improves the efficiency of file management using file abstraction and sharing between the real and virtual storage centers in large-scale networks. The security in storage-centric system aims at resilience to security threats and prevents unauthorized access to sensitive and confidential information. Storage dimensions vary in PAN and LAN systems. PAN and WBAN storage systems consist of the low payload and reduced header formats. Further extension of SCN includes the elucidation of storage and management techniques in heterogeneous and edge computing systems for smart environments.

References

1. Newsome J. and Song D., "GEM: Graph EMbedding for Routing and DataCentric Storage in Sensor Networks without Geographic Information," In *ACM 1st International Conference on Embedded Networked Sensor Systems (SenSys)*, Los Angeles, CA, 5–7 November 2003, pp. 76–88.
2. Zhao W., Ammar M., and Zegura E., "A Message Ferrying Approach for Data Delivery in Sparse Mobile Ad Hoc Networks," In *ACM 5th International Symposium on Mobile Ad Hoc Networking and Computing (MobiHoc)*, Roppongi Hills, Tokyo, Japan, 24–26 May 2004, pp. 187–198.
3. Diao Y., Ganesan D., Mathur G., and Shenoy P., "Rethinking Data Management for Storage-Centric Sensor Networks," In *Third Biennial Conference on Innovative Data Systems Research (CIDR)*, Asilomar, CA, USA, 7–10 January 2007, pp. 22–31.
4. Luo L., Huang C., Abdelzaher T., and Stankovic J., "EnviroStore: A Cooperative Storage System for Disconnected Operation in Sensor Networks," In *IEEE 26th International Conference on Computer Communications (INFOCOM)*, Barcelona, Spain, 6–12 May 2007, pp. 1802–1810.
5. Girao J., Westhoff D., Mykletun E., and Araki T., "TinyPEDS: Tiny Persistent Encrypted Data Storage in Asynchronous Wireless Sensor Networks," *Elsevier Journal of Ad Hoc Networks*, Vol. 5(7), September 2007, pp. 1073–1089.
6. Luo H., Xing G., Li M., and Jia X., "Dynamic Multi-Resolution Data Dissemination in Storage-Centric Wireless Sensor Networks," In *ACM 10th Symposium on Modeling, Analysis, and Simulation of Wireless and Mobile Systems (MSWiM)*, Chania, Crete Island, Greece, 22–26 October 2007, pp. 78–85.

7. Singh A., Korupolu M., and Mohapatra D., "Server-Storage Virtualization: Integration and Load Balancing in Data Centers," In *ACM/IEEE Conference on Supercomputing (SC)*, Austin, TX, 15–21 November 2008, pp. 53:1–53:12.

8. Nguyen N., Krishnamurthy S., Xie P., and Jones D., "Improving Information Availability in Storage-Centric Sensor Networks," In *IEEE 33rd Conference on Local Computer Networks (LCN)*, Montreal, Quebec, Canada, 14–17 October 2008, pp. 82–90.

9. Zeng W., Zhao Y., Ou K., and Song W., "Research on Cloud Storage Architecture and Key Technologies," In *ACM 2nd International Conference on Computers and Information Systems (ICIS)*, Seoul, Korea, 24–26 November 2009, pp. 1044–1048.

10. Wang L., Noh D. K., Yang Y., Le H. K., Abdelzaher T. F., and Ward M., "AdaptSens: An Adaptive Data Collection and Storage Service for Solar-Powered Sensor networks," In *IEEE 30th Real-Time Systems Symposium*, Washington, DC, 1–4 December 2009, pp. 303–312.

11. Moga A., Botan I., and Tatbul N., "UpStream: Storage-Centric Load Management for Streaming Applications with Update Semantics," *Springer Journal on Very Large Data Bases*, Vol. 20(6), 1 December 2011, pp. 867–892.

12. Cuevas A., Uruena M., Veciana G. D., Cuevas R., and Crespi N., "Dynamic Data-Centric Storage for Long-Term Storage in Wireless Sensor and Actor Networks," *Springer Journal of Wireless Networks*, Vol. 20(1), 1 January 2014, pp. 141–153.

13. Luo T., Ma S., Lee R., Zhang X., Liu D., and Zhou L., "S-CAVE: Effective SSD Caching to Improve Virtual Machine Storage Performance," In *ACM 22nd International Conference on Parallel Architectures and Compilation Techniques (PACT)*, Edinburgh, UK, 7 October 2013, pp. 103–112.

14. Jiang L., Xu L. D., Cai H., Jiang Z., Bu F., and Xu B., "An IoT-Oriented Data Storage Framework in Cloud Computing Platform," *IEEE Transactions on Industrial Informatics*, Vol. 10(2), May 2014, pp. 1443–1451.

15. Jiang H., Shen F., Chen S., Li K.-C., and Jeong Y.-S., "A Secure and Scalable Storage System for Aggregate Data in IoT," *Elsevier Journal of Future Generation Computer Systems*, Vol. 49, August 2015, pp. 133–141.

16. Sourlas V., Flegkas P., Paschos G. S., Katsaros D., and Tassiulas L., "Storage Planning and Replica Assignment in Content-Centric Publish/Subscribe Networks," *Elsevier Journal of Computer Networks*, Vol. 55(18), 29 December 2011, pp. 4021–4032.

17. Maenhaut P.-J., Moens H., Volckaert B., Ongenae V., and Turck F. D., "Design of a Hierarchical Software-Defined Storage System for Data-Intensive Multi-Tenant Cloud Applications," In *IEEE 11th International Conference on Network and Service Management (CNSM)*, Barcelona, Spain, 9–13 November 2015, pp. 22–28.

18. Wang B., Li B., and Li H., "Panda: Public Auditing for Shared Data with Efficient User Revocation in the Cloud," *IEEE Transactions on Services Computing*, Vol. 8(1), January–February 2015, pp. 92–106.

19. Gracia-Tinedo R., Garcia-Lopez P., Sanchez-Artigas M., Sampe J., Moatti Y., Rom E., Naor D., Nou R., Cortes T., Oppermann W., and Michiardi P., "IOStack: Software-Defined Object Storage," *IEEE Internet Computing Magazine*, Vol. 20(3), May–June 2016, pp. 10–18.

20. Bagci I. E., Raza S., Chung T., Roedig U., and Voigt T., "Combined Secure Storage and Communication for the Internet of Things," In *IEEE 10th Annual Communications Society Conference on Sensor, Mesh and Ad Hoc Communications and Networks (SECON)*, New Orleans, LA, 24–27 June 2013, pp. 523–531.

21. Huang Z., Chen J., Lin Y., You P., and Peng Y., "Minimizing Data Redundancy for High Reliable Cloud Storage Systems," *Elsevier Journal of Computer Networks*, Vol. 81, 22 April 2015, pp. 164–177.

22. Xiao Z., Song W., and Chen Q., "Dynamic Resource Allocation Using Virtual Machines for Cloud Computing Environment," *IEEE Transactions on Parallel and Distributed Systems*, Vol. 24(6), June 2013, pp. 1107–1117.

23. Spillner J., Muller J., and Schill A., "Creating Optimal Cloud Storage Systems," *Elsevier Journal of Future Generation Computer Systems*, Vol. 29(4), June 2013, pp. 1062–1072.

24. Shen J., Zhou T., Chen X., Li J., and Susilo W., "Anonymous and Traceable Group Data Sharing in Cloud Computing," *IEEE Transactions on Information Forensics and Security*, Vol. 13(4), April 2018, pp. 912–925.

25. Mottola L., "Programming Storage-Centric Sensor Networks with Squirrel," In *ACM/IEEE 9th International Conference on Information Processing in Sensor Networks (IPSN)*, Stockholm, Sweden, 12–16 April 2010, pp. 1–12.

26. Dudkowski D. and Hahner J., "An Analytical Study of the Communication Cost of Data-Centric Storage in Mobile Ad Hoc Networks," *Elsevier Journal of Ad Hoc Networks*, Vol. 10(3), May 2012, pp. 356–372.

27. Carofiglio G., Gallo M., Muscariello L., and Perino D., "Evaluating Per-Application Storage Management in Content-Centric Networks," *Elsevier Journal of Computer Communications*, Vol. 36(7), 1 April 2013, pp. 750–757.

28. Xu Q., Arumugam R. V., Yong K. L., and Mahadevan S., "DROP: Facilitating Distributed Metadata Management in EB-scale Storage Systems," In *IEEE 29th Symposium on Mass Storage Systems and Technologies (MSST)*, Long Beach, CA, 6–10 May 2013, pp. 1–10.

29. Xue X., Hou X., Tang B., and Bagai R., "Data Preservation in Intermittently Connected Sensor Networks with Data Priority," In *IEEE International Conference on Sensing, Communications and Networking (SECON)*, New Orleans, LA, 24–27 June 2013, pp. 122–130.

30. Xu M., Song W.-Z., Heo D., Kim J.-H., and Kim B.-S., "ECPC: Preserve Downtime Data Persistence in Disruptive Sensor Networks," In *IEEE 10th International Conference on Mobile Ad-Hoc and Sensor Systems (MASS)*, Hangzhou, China, 14–16 October 2013, pp. 281–289.

31. Gonizzi P., Ferrari G., Gay V., and Leguay J., "Data Dissemination Scheme for Distributed Storage for IoT Observation Systems at Large Scale," *Elsevier Journal of Information Fusion*, Vol. 22, March 2015, pp. 16–25.

32. Shi B. and Srivastava A., "Thermal and Power-Aware Task Scheduling for Hadoop Based Storage-centric Datacenters," In *IEEE International Green Computing Conference (IGCC)*, Chicago, IL, 15–18 August 2010, pp. 73–83.

33. Mesnier M., Akers J. B., Chen F., and Luo T., "Differentiated Storage Services," In *ACM 23rd Symposium on Operating Systems Principles (SOSP)*, Cascais, Portugal, 23–26 October 2011, pp. 57–70.

34. Talari A. and Rahnavard N., "CStorage: Distributed Data Storage in Wireless Sensor Networks Employing Compressive Sensing," In *IEEE Global Telecommunications Conference (GLOBECOM)*, Houston, TX, 5–9 December 2011, pp. 1–5.

35. Ren J. and Yang Q., "I-CASH: Intelligently Coupled Array of SSD and HDD," In *IEEE 17th International Symposium on High Performance Computer Architecture (HPCA)*, San Antonio, TX, 12–16 February 2011, pp. 278–289.

36. Su M., Zhang L., Wu Y., Chen K., and Li K., "Systematic Data Placement Optimization in Multi-Cloud Storage for Complex Requirements," *IEEE Transactions on Computers*, Vol. 65(6), 1 June 2016, pp. 1964–1977.

37. Kleineweber C., Reinefeld A., and Schutt T., "QoS-Aware Storage Virtualization for Cloud File Systems," In *ACM 1st International Workshop on Programmable file systems (PFSW)*, Vancouver, BC, 23–27 June 2014, pp. 19–26.

38. Wang Q., Ren K., Yu S., and Lou W., "Dependable and Secure Sensor Data Storage with Dynamic Integrity Assurance," *ACM Transactions on Sensor Networks (TOSN)*, Vol. 8(1), August 2011, pp. 9:1–9:24.

39. Li M., Lou W., and Ren K., "Data Security and Privacy in Wireless Body Area Networks," *IEEE Wireless Communications Magazine*, Vol. 17(1), February 2010, pp. 51–58.

40. Sheng B. and Li Q., "Verifiable Privacy-Preserving Sensor Network Storage for Range Query," *IEEE Transactions on Mobile Computing*, Vol. 10(9), September 2011, pp. 1312–1326.

41. Chen F., Xiang T., Yang Y., and Chow S. S. M., "Secure Cloud Storage Meets with Secure Network Coding," *IEEE Transactions on Computers*, Vol. 65(6), 1 June 2016, pp. 1936–1948.

42. Cuevas A. and Bouckerche A., "Improving Security in Data-Centric Storage for Wireless Sensor Networks," In *ACM 10th Symposium on Performance Evaluation of Wireless Ad Hoc, Sensor, & Ubiquitous Networks (PE-WASUN)*, Barcelona, Spain, 3–8 November 2013, pp. 129–136.

43. Snyder P. and Kanich C., "Cloudsweeper and Data-Centric Security," *ACM Newsletter SIGCAS Computers and Society*, Vol. 44(2), July 2014, pp. 8–10.

44. Li S., Cao Q., Wan S., Qian L., and Xie C., "HRSPC: A Hybrid Redundancy Scheme via Exploring Computational Locality to Support Fast Recovery and High Reliability in Distributed Storage Systems," *Elsevier Journal of Network and Computer Applications*, Vol. 66, May 2016, pp. 52–63.

45. Xue Y., Li J., Nazarian S., and Bogdan P., "Fundamental Challenges Toward Making the IoT a Reachable Reality: A Model-Centric Investigation," *ACM Transactions on Design Automation of Electronic Systems (TODAES)*, Vol. 22(3), May 2017, pp. 53:1–53:25.

46. Fan R., Ping L.-D., Fu J.-Q., and Pan X.-Z., "The New Secure and Efficient Data Storage Approaches for Wireless Body Area Networks," In *IEEE International Conference on Wireless Communications and Signal Processing (WCSP)*, Suzhou, China, 21–23 October 2010, pp. 1–5.

47. Al-Kiswany S., Costa L. B., Yang H., Vairavanathan E., and Ripeanu M., "A Cross-Layer Optimized Storage System for Workflow Applications," *Elsevier Journal of Future Generation Computer Systems*, Vol. 75, October 2017, pp. 423–437.

48. Pedersen J., Amat A. G. I., Andriyanova I., and Brannstrom F., "Distributed Storage in Mobile Wireless Networks with Device-to-Device Communication," *IEEE Transactions on Communications*, Vol. 64(11), November 2016, pp. 4862–4878.

49. Sengupta A., Tandon R., and Simeone O., "Cache Aided Wireless Networks: Tradeoffs between Storage and Latency," In *IEEE Annual Conference on Information Science and Systems (CISS)*, Princeton, NJ, 16–18 March 2016, pp. 320–325.

6

CLOUD-CENTRIC NETWORKING USING VIRTUALIZATION AND RESOURCE SHARING

M. BALA KRISHNA, DANIEL CORUJO, DIOGO GOMES, JOÃO PAULO BARRACA AND HELDER MOREIRA

Contents

6.1 Cloud-Centric Networking

Cloud Computing is today's dominant computing paradigm which companies, governments and individuals rely upon to meet their computation, storage and communication needs. Applications living in the Cloud pose several new challenges to networks, not only due to their needs for transferring massive amounts of data to and from the data centers where the Cloud infrastructure is deployed, but mostly due to the elasticity and dynamics of the services being offered. In this section we address the most relevant research work in the area of Cloud-Centric Networking, their advantages and shortcomings, as well as potential leads on new innovations. We define Cloud-Centric Networking (CCN) as the paradigm by which all network communications occur in the Cloud, encompassing a very broad list of use cases from Web and Mobile Applications Backend to Network Function Virtualization.

Cloud Computing solutions are usually centrally hosted in remote data centers where millions of servers are interconnected, typically using a tree-shaped architecture composed of two or three levels of deeply interconnected switches [1]. At the top level, a core tier roots all the branches while servers are placed at the bottom as leaves. These topologies can support up to 8K physical hosts and are limited by the upstream bandwidth and switching capacity of the switches used to aggregate hosts towards the core tier. This solution is nonetheless quite expensive, especially considering the top-most switches, and is usually the main bottleneck of applications running on this infrastructure. As applications and services become increasingly more Cloud Centric, it is foreseen that the amount of traffic exchanged inside data centers will increase. The growth in Mobile Computing [2], which uses the Cloud for most of its computational and storage needs, will also require more from the network fabric of the data center.

The Cloud is defined by its elasticity, the capability to scale and free resources upon request. Server virtualization techniques have provided the means to achieve such elasticity, as Virtual Machines (VMs) can be launched and terminated in very short timeframes. Network virtualization has also enabled Cloud users to create and delete networks upon request, allocating networking resources as needed while providing increased security levels through a better segmentation of tenants [3].

It is nonetheless important to consider the overhead created by the VMs on physical network devices. Switch forwarding tables increase in size due to the amount of VMs, packet overhead increases due to virtualization techniques such as VxLAN, IP GRE and others with important impacts in packet fragmentation and useful data transfer.

Infrastructure-as-a-Service (IaaS), Platform-as-a-Service (PaaS) and Software-as-a-Service (SaaS) [4] models correlate to Cloud-Centric Networking at different layers. In an IaaS model the network is yet another service that can be requested by tenants, with resources for such networks able to be reserved and managed on demand. In the PaaS model, networks are abstracted for the tenant and network resources are centrally managed at scale by the Cloud provider, even though the tenant can allocate various network services. In SaaS, the network is completely opaque to the tenant and all network aspects are managed by the provider, with the tenants merely consuming services at the edge of the Cloud.

Software-defined Networking (SDN) [5] is currently seen by both academia and industry as the necessary paradigm shift to increase the flexibility of networks in a Cloud-Centric Networking environment. Automation of network creation, deployment and management as well as flexibility and elasticity of network resources and Quality of Service (QoS) are vital to the deployment of complex applications in a Cloud environment. SDN enables the separation of the control and data planes, allowing networks to be managed and programmed by external applications through Application Programming Interfaces (APIs) [6]. The data plane, which remains in dedicated hardware (network devices), combined with the control plane which consists of a software controller, provides a scalable and flexible platform for deploying network applications. This allows for an efficient, centralized and holistic network management while having a reduced operating cost, which are the main advantages of using SDN in a Cloud environment. Network Function Virtualization (NFV) shares some of the base concepts of SDN but goes further, as all network elements and functions are virtualized. It consists of creating virtualized network domains on top of virtual software components that are able to run independently from the underlying hardware [7]. While sharing most of the SDN advantages, NFV adds some more benefits to CCN. First, there is a reduced equipment cost and energy consumption because the same hardware used to provide VMs may be utilized to create the virtual networks. Then, there is even more efficiency, scalability, flexibility and elasticity inherited from the existing dynamic resource allocation in the Cloud Computing area [8,9].

The following sections will further explore the topic of CCN in regards to multi-Cloud networking (Section 6.2), virtualization (Section 6.3), resource sharing and mobility models (Section 6.4) and, finally, security (Section 6.5).

6.2 Multi-Cloud Networking

There has been a clear evolution towards the integration of systems into richer environments, potentially exploiting locality, and heterogeneity in both Cloud offers and product demand, as well as the opportunities provided by the different legal and technological frameworks. In this sense, a single Cloud provider can frequently be incapable of providing (or hosting, depending on the nature of the service provider) an entire set of services, for all the potential clients. Even if it has the capacity, and if it fits within the typical latency and legal constraints for services, several aspects such as availability, locality (associated with both latency, legality and cost) or redundancy (among others) are major drivers towards a distributed architecture leveraging multiple Clouds. The concept of a united Cloud was coined as the Intercloud, which is related to Multi-Cloud. After an initial set of solutions, the concepts for multiple Clouds are maturing and more robust solutions, which also consider Cloud providers, are being proposed [10]. The telecom world also moved towards integrating the Cloud with their networks [11], as well as in the form of Mobile Edge Computing (MEC), with latency as the main driver, especially under the scope of 3G-PPP 5G [12], and the strict requirements imposed by future 5G network infrastructure. In this vision, a MEC-enabled telco network is composed by several cloudlets, located at the access fringe, and managed in a centralized manner. Intercloud, or more simply Multi-Cloud, considers that network management concepts are of major importance in order to provide a unified communication substrate for applications, in particular when dealing with dynamic Cloud environments in telco networks [13]. At the scale of the Internet, and outside each telco world, businesses are evolving to

Multi-Cloud networks, or already provide services over established Multi-Cloud networking architectures, even if most strategies rely on simple integration of the platforms using overlays, and there is no strict integration of the Cloud networking substrate [14], [15]. Some solutions envision partial collaboration [16], whereas some already available products provide Multi-Cloud orchestration of resources, both for end-clients and Cloud providers.

In the context of users of public Clouds, access to the networking fabric is not readily available. Therefore, interconnection of networking worlds from different Cloud providers is possible, and presents benefits, but it relies on the existence of brokers, common managers or orchestrators. Network elements are standard virtual machines, and networking between Clouds is achieved through overlays. One of the initial solutions for an interconnected Cloud was presented by the IEEE Intercloud Testbed [17] and focused on a scenario where players have reasonable resources, and develop a solution that is adaptable to a multitude of Cloud providers. This approach also mimicked the existing knowledge about the operation of the Internet, proposing the existence of Intercloud Exchange points (with some similarity to Internet Exchange Points) and acting as aggregators of traffic between providers. Also, (software) routers, mimicking Autonomous Systems Edge Routers in the Border Gateway Protocol (BGP), provided transparent connectivity. Without strictly adopting NFV [18] concepts, and with a focus on federation and higher layer services, this was an important step towards a software-defined Multi-Cloud networking environment, aiming for an integrated Cloud environment. With the advent of SDN [19] and NFV, the Multi-Cloud approach, which explicitly exploits multiple Clouds even if not aiming for a single unifying Cloud, took a different direction for its clients. The enhanced flexibility provided by SDN made it much faster to deploy and re-arrange Virtual Network Functions (VNF) in IaaS providers, and even some Cloud providers already provide VNFs as a service, potentially operating as Network Function Virtualization Infrastructure (NFVI) providers. As the Clouds matured, as well as with the demonstrated need of Multi-Cloud networking provided by SDN and the concepts related to slicing, challenges were related to how to integrate and then orchestrate heterogeneous, and loosely coupled, resources across multiple management domains.

6.2.1 Orchestration

Having multiple Clouds presents potential capacity, which requires the existence of coordinated management mechanisms in order to actually exploit that capacity. Orchestrators tap into the available resources of a Cloud provider, and take in consideration the existing demands to coordinate how those resources are used in the implementation of a particular service or function. Standardization bodies are addressing these challenges, leading the community towards orchestrators that tend to be provider agnostic, and reusable. Researchers and the industry are developing specific solutions, that mostly address Future Internet scenarios or the work developed at the standardization bodies. The ETSI Management and Orchestration (MANO) group is one of the most relevant groups, aiming to develop an open source orchestration solution (Open Source MANO).[1] Aligned with the work being developed at the ETSI NFV group (that addresses the lifecycle, creation and interfaces of an VNF), it creates a Cloud agnostic solution called a Network Function Virtualization Orchestrator (NFVO) [20–22] as a building block for telecom providers and other Cloud users. It also builds upon other solutions related to virtualization, SDN and NFV, such as Linux KVM, Open vSwitch and OpenDayLight with a focus towards OpenStack Heat (as the reference open source virtualization component). This work has been extremely relevant and it is a building block for many other solutions, such as the case of T-NOVA [23].

While the NFVO has a complete view over the networking environment, placement of VNFs and the interconnecting ports, it still relies on the support of each Cloud provider to implement the desired orchestration policies. Virtualized Infrastructure Managers (VIMs) keep inventories of virtual and physical resources at each Cloud and deal with the effective instantiation of the VNFs in the Cloud environment they manage. As the VIMs are required only to implement a specific behavior, communicating through well-defined interfaces, other players can develop VIMs (e.g., VMware vSphere) for their Cloud environments, enabling a multitude of solutions to be integrated under the same NFVO. End-to-end orchestration is also considered

---

[1] http://www.etsi.org/technologies-clusters/technologies/nfv/open-source-mano

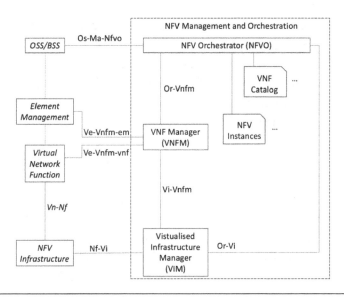

Figure 6.1   ETSI NFV-MANO overall architecture.

in MANO (E2E MANO) by a specific management plane dealing with inter-operator service level agreement fulfillment, quality of service, and contracts that also span multiple domains and multiple operators. Still, the business logic is limited in ETSI MANO as this is not part of the group's scope. Figure 6.1 depicts the ETSI MANO architecture, and its main modules.

Other standardization bodies also took the challenge of utilizing Multi-Cloud networks, with IEEE focusing on the Next Generation Service Overlay Network (NGSON) [24] in P1903. NGSON arises from the world of IP-based telecom providers, which adopted the IP Multimedia Subsystem (IMS) and operate a network based on somewhat rigid interfaces and high fragmentation of functions. In this scenario, Average revenue per user (ARPU) has been steadily declining, and providers are aiming for higher integration of their networks, with higher engagement with their clients. The solution is focused on services that are provided across multiple Clouds, but are presented as being in a single networking environment. The envisioned functions encompass the entire spectrum of multidomain product delivery. Service functions provide the means for service registry, and discovery, as well as routing and composition. Transport functions create an overlay network with advanced mechanisms for routing optimization, dynamic policy enforcement and in particular context-aware routing which is able to exploit locality and slicing. Management and operation functions deal with service lifecycle, QoS and Service Level Agreements (SLAs). A key aspect of the multidomain communication solution is the NGSON Service Routers, which enable the overlay at boundary points of the telco network, and effectively interconnect different sites.

In an alternative effort to develop yet another solution for orchestration or service provisioning, ITU is focusing their efforts on a more holistic vision of future networks, where SDN and Multi-Cloud networking are a core function [22]. In this sense, ITU addresses the possible functions to be deployed in Multi-Cloud networks, which are not limited to network functions, but also include scenarios such as interconnected Content Delivery Networks (CDNs) and orchestration in Next Generation Network Elements (NGNe), Internet of Things or block chains.

T-NOVA [23] is another reference work, which builds upon the interests and needs of the partners belonging to the T-NOVA consortium and is aligned with ETSI NFV [18] and ETSI MANO. It can be seen as an instantiation of ETSI MANO in a business environment, with a clear focus towards service provisioning and actual business operation, in a Multi-Cloud and multioperator networking environment. T-NOVA considers the addition of modules that exploit the ETSI MANO architecture for general orchestration of resources, but adds functionalities related to billing, AAA and SLA Management, which are vital for monetization of an integrated Multi-Cloud. In T-NOVA, the orchestrator components also consider the existence of a marketplace for VNFs, exposed by VNF catalogs, which providers can use to implement richer networks or application-driven slices. This also enables new players to develop solutions for specific markets, and make VNFs rapidly available to a large base of potential clients.

An interesting aspect of T-NOVA is the addition of mechanisms for automatic SLA evaluation and the definition of the respective penalties and rewards. With the increased dynamics and reduced coupling of Multi-Cloud networking, these functions are vital for the provisioning of services with predictable QoE. Much of the work in this area was inspired by a previous initiative, FP7 ETICS [25], which focused in the existing Internet structure and telecom operator management, considering the issues arising from the dynamic interconnection of telecom providers, and the existence of services that span multiple domains. In particular, it considered the challenges of the end-to-end provisioning of complex services, composed by sub-services existing at different locations, or provided to third-party networks, with predictable QoS and QoE guarantees.

6.2.2 *Optimal Network Function Placement*

One of the most basic challenges for Multi-Cloud networking is VNF placement across the multiple providers, and this is a basic function for orchestrators. Of particular interest is the aspect of Service Function Chaining (SFC) across providers, or how to chain the multiple VNFs that compose a service (or a slice) in such a way that the service is actually provided in an optimal way, but across multiple providers. The issues arising from this functionality are the same as with traditional SFC inside a single provider, but now with the challenges of establishing and monitoring SLAs across providers, and the added difficulty of providing an expected QoE, with an higher layer of heterogeneity in the underlying network. Without effective solutions addressing this issue, orchestration is technically useful, but no added value can be extracted. Yet, SDN is now viewed as a service, which provides a networking layer that is abstracted from the actual networking fabric, following service-oriented concepts [26,27]. *Composition* is the key term, but now it should be considered that the networking fabric can place limits on and shape the final composition.

Solutions must focus on the deployment of services over Multi-Cloud environments, while envisioning the maximization of some performance-oriented goals such as capacity [28], Quality of Service [29] or optimal server location based on performance or topological distance [30,31].

6.3 Virtualization in CCN

As we witness an increasing reliance on Cloud-based deployments, not only within the ICT sector but also to support other societal areas, virtualization aspects in CCN pertain to two major directions in regards to its applicability. On one hand, as data centers grow, more sophisticated measures for controlling and managing information interexchange between associated elements is required. On the other, the demands of the virtualized applications and services running within the Cloud-Centric Network see their requirements increase, due to the explosion of on-line and mobile access from users.

[5] evidences a core set of the most common reasons for resource virtualization, complemented by [32], namely

- *Sharing*: To take advantage of under-used resources, by exploiting appropriate techniques (i.e., resource multitasking, sharing or prioritization) that allow more tasks to globally use available resources in an optimal way;
- *Isolation*: To securely separate the usage space of different users and/or tasks sharing the same resources;
- *Aggregation*: To orchestrate and combine multiple resources allowing the platform to fulfill tasks with greater demands;
- *Dynamics (Scalability and Flexibility)*: To allow virtually deployed resources to rapidly adapt to changing requirements and conditions, independently of the specifications of the underlying physical network;
- *Ease of Management*: To provide software-based control and management approaches for resource deployment and configuration, instead of physical hardware operations;
- *Recursion*: Also known as *nesting*, which considers multiple tiers of virtual networks on top of one another, with the lower tier being the only one composed by physical entities;
- *Inheritance*: When top-tier virtual networks inherit properties belonging to their lower tier counterparts;
- *Stability and Convergence*: Preventing configuration errors from the physical network to manifest in the virtual substrate;
- *Heterogeneity*: When not only the physical network but also the virtual networks on top of it can be comprised of different technologies and protocols.

In this sense, it is critical that Cloud-Centric Networking operations become supported by virtualization-based mechanisms that maintain such principles, focusing on the provisioning of high performance platforms that are able to be dynamically deployed in different entities of a (virtualized) network, with low cost.

The following section addresses current virtualization aspects and approaches in the CCN domain.

6.3.1 Virtualization for CCN Operations

By supporting the orchestration and management dynamics of virtualized Cloud-based services provisioning requires flexible data flow switching and routing support from the infrastructure, users have the ability to create, on demand, new virtual machines and virtual networks to connect them. According to [1], early large cluster-based solutions either favored specialized communication protocols and hardware (i.e., InfiniBand or Myrinet), or, instead, the usage of commodity Ethernet switching and routing hardware for linking the cluster nodes. In the first case, greater scalability was achieved at a higher cost and with no native compatibility with TCP/IP services and applications. In the second case, although more cost efficient and supporting existing applications, hardware and operating systems, its bandwidth scalability possibilities incurred greater cost.

This, of course, attracted the development of proposals targeting the provisioning of virtualized networks interconnecting the different virtual machines (usually belonging to different users) existing in physical machines. Such proposals aimed to evolve traditional network virtualization techniques, namely Virtual Local Area Networks (VLANs) and virtual private networks (VPN), which lacked the network decoupling from the underlying physical infrastructure. In this way, solutions such as Virtual Extensible LANs (VxLAN) by VMWare joined Network Virtualization using Generic Routing ENcapsulation (NVGRE) and Stateless Transport Tunneling (STT) as protocols for this effect. Other solutions, as in [33], leverage other traditional virtualization techniques such as overlay networks using tunnels, and actually provide *virtual* overlay networks (VON) that make use of virtual switches that are located on the edges of a data center network. This provides a comprehensive protocol-agnostic network fabric, which can be configured for service provisioning without requiring manual configuration of the involved core network elements. This contributes to a CCN virtualization model where the network intelligence is realized via software that resides at the edge in virtual switches, which precludes the requirement of having to realize changes in the actual core infrastructure itself.

Router virtualization solutions, such as [34], already provided programmable forwarding plane configuration capabilities able to deliver high performance due to their reliance on optimized core scheduling and multi-queuing for network interface sharing, while still being able to be deployed using commodity hardware. Other approaches, such as Open Application Delivery Network (OpenADN) [5], allowed providers to establish application-level policies concerning their exchanged messages and packets.

The different platforms resulting from these research efforts evidenced as well the opportunity to develop advances at the hardware level, allowing higher throughput and lower latency in packet processing for virtual-based operations. For example, according to [35], direct access to I/O interfaces is preferable for lower latency and higher throughput. Concretely, at the network interface card, Intel provided a Data Plane Development Kit (DPDK[2]), lowering packet processing overhead by allowing user applications to receive data directly from the interface (which was a detriment for different virtual machines connected to the same physical network interface). Evolutions such as NetVM [36] evolved from the reduced set of virtualization support options from DPDK, allowing for more complex network operations at line speeds greater than 10 Gbps, while maintaining the ability for network services to be created, chained and balanced in a flexible way. This capability was achieved by using shared memory and an improved CPU scheduler, supported by capabilities such as single-root I/O virtualization (SR-IOV) compliance, which offer faster and higher throughput to virtual machines, when compared to interfaces managed by the hypervisor (albeit at the expense of more complex configuration and support of live virtual machine migration).

As an outcome, we are approaching the concept of Network as a Service (NaaS) [33], where Cloud computing and virtualisation are provided to users and operators alike, following an on-demand and self-service provisioning model. Central to this model are the roles of Infrastructure Provider (InP) and Service Provider (SP) [32], where the former allocates a part of its physical infrastructure resources to the latter by embedding a virtual network on one or more physical

---

[2] Data Plane Development Kit, http://dpdk.org

networks, thus allowing the SP to focus on the provisioning of end-to-end services to users.

6.3.2 *Virtualization Enhancements in CCN-Supported Services*

ETSI has presented a set of relevant use cases for the application of virtualized network services, which was summarized in [37] as follows:

- *Network Functions Virtualization as a service:* Occurs when telecommunications providers supply NFV elements (ranging from the whole platform to a single virtualized network function instance) as a service;
- *Virtualization of Mobile Core Network and IMS:* When the mobile core network functions are provided by virtualized counterparts, aiming to reduce costs associated with hardware and increase flexibility. It allows a flexible distribution of hardware resources, eliminates hardware bottlenecks and allows rapid launch of innovative services;
- *Virtualization of Mobile Base Station:* Separating signal processing from the radio access hardware, and combining it in a single hardware platform for multiple base stations;
- *Virtualization of the Home Environment:* Reducing maintenance costs associated with the installation and/or configuration of new equipment;
- *Virtualization of CDNs:* Optimizing caching resources according to service usage;
- *Fixed Access Network Functions Virtualization:* Supporting multi-tenancy in access network devices and partition access nodes.

One of the main concerns when service providers turn their business provision from physical middle boxes to a Cloud-based deployment is the risk that a virtualized infrastructure can have in regards to reaching the desired levels of service reliability, availability and quality demanded by the applications and services [38]. Specifically, [39] emphasizes that virtualization can impose requirements on the network's ability to provide mobility, scaling and isolation, to a greater degree when compared with physical counterparts [40]. For example, as evidenced in [41], studies exist (e.g., the case of Amazon EC2 Data

Center) that verified latency and throughput instability despite low utilization of the underlying network. This aspect initiated a large research and experimentation effort (even before the advent of 5G) undergone by projects such as the EU FP7 Mobile Cloud Networking (MCN[3]), where new advanced techniques for the coordination of distributed Cloud resources, network function virtualization and other important network-based management aspects (e.g., load balancing and mobility management) were more tightly integrated with the necessary orchestration procedures in the underlying virtualized infrastructure. [42] also highlighted other important projects, namely i) iJOIN[4] which pursued the concept of RAN-as-a-Service (RANaaS) and optimized both access and backhaul in heterogeneous environments; ii) MODAClouds[5] which addressed end user service portability to the Cloud and iii) CloudScale[6] which focused on the prediction, analysis and solution of Cloud-based deployed services scalability issues. More recently, in the advent of 5G networks, projects such as T-NOVA[7] addressed goals similar to MCN, but based on NFV mechanisms.

[43] highlights that, in several deployments (e.g., Amazon), the number of virtual ports has surpassed the number of physical ones. This acknowledges the demands for more flexible means of virtualized networks deployment, evolving from the end-to-end principle into hypervisor-assisted overlay communications, managed by SDN controllers. Nonetheless, the reliance on virtualized service deployment for network-based operations is also able to provide aspects that are unavailable directly for physical deployments. Concretely, as evidenced by [39], the usage of virtualized settings configuration by hypervisors allows them to be aware of host events, which allows the virtual networking layer to directly learn that information and use it to better manage configurations and the forwarding state of virtualized network elements.

More recently, the integration of SDN and NFV into network architectures allowed for more simple, dynamic and flexible network

---

[3] Mobile Cloud Networking, EU FP7 Project, http://www.mobile-Cloud-networking.eu/

[4] iJOIN, EU FP7 Project, http://www.ict-ijoin.eu

[5] MODAClouds, EU FP7 Project, http://www.modaclouds.eu/project/funding/

[6] CloudScale, EU FP7 Project, http://www.cloudscale-project.eu

[7] T-NOVA, EU H2020 Project, http://www.t-nova.eu

control and management to support services running in Cloud-based infrastructures. The combined operation of both SDN and NFV contributes to a reduction of hardware equipment configuration and associated networking operations, and allows for better network resources utilization by means of jointly using network virtualization and centralized intelligence [44]. The application of SDN and NFV in these environments raises new challenges in the form of resource allocation, particularly in the case of hybrid scenarios simultaneously featuring both virtual and physical resources. For example, although it is straightforward to instantiate some network functions in a virtual way, other functions have stricter delay requirements or topology dependencies, or they need to be directly on the datapath of two intervening endpoints. These are further exacerbated considering that VNFs should be instantiated at appropriate locations (i.e., considering closeness to information generation and consumption, as well as the availability of resources, or even the case of data traffic starting to traverse indirect paths due to the placement of VNFs in the data center), as well as being able to dynamically allocate and scale hardware resources as needed, or even chaining them to promote their interconnectivity [45]. A final, but important, aspect is related to the requirement for coexistence between VNFs and legacy networks, considering deployment purposes, namely preventing service disruption motivated by the transition of a network service from a physical entity to a virtualized one, as well as avoiding forwarding graph inconsistencies motivated by the existence of more than one VNF.

As the solutions and outcomes of virtualized networks start to see an increased deployment, it is possible to glimpse novel directions for CCN virtualisation. Currently, the Next Generation Mobile Network Alliance (NGMN[8]) and the 5G-PPP[9] have envisioned network slicing prospects, ranging from its application at the core to the provisioning of end-to-end slices [46]. Network slicing [41] sees the individual physical network divided into multiple isolated logical networks, each capable of fulfilling different requirements, supported by infrastructure virtualization mechanisms. The provisioning of slices goes beyond full isolation of resources, but actually demands the

---

[8] Next Generation Mobile Network Alliance, https://www.ngmn.org
[9] 5G Infrastructure Public Private Partnership, https://5g-ppp.eu

ability to have different control operations supported over the different slices (and equally, by the underlying virtualization substrate that provides them). In this sense, network slicing distances itself from existing network virtualization solutions, by opening up the possibility of allowing third parties to fully take charge of a slice control. Nonetheless, the realization of this vision is still underway, particularly in its end-to-end aspect, where the application of slicing strategies at the RAN is still at an early stage.

6.4 Resource Sharing and Mobility Models in CCN

Virtualization not only is able to increase networking programmability features but is also capable of providing important supportive measures such as VM cloning and mobility, which support secure and efficient physical resource sharing by enabling VM snapshots to be taken and deployed into underutilized physical nodes [43]. These are important requirements for an ever-increasing set of businesses that aim to incorporate Cloud-based network virtualization mechanisms into their core operations, such as mobile network operators (MNO), and even offload computing tasks from mobile devices in order to allow applications running therein to tap into more powerful resources [47] (something also realized at network function level, identified as Cloud bursting [14]), which are able to contribute to the growing vision of a Cloud-supported Internet of Things [48]. In addition, infrastructure sharing between operators is also a promising solution towards cost reduction, and can reach all the way to the sharing of not only computing and storage, but also spectrum sharing [49].

However, these are still regarded as a challenge for Cloud computing [42], and they include a myriad of very specialized components and services, which need to be supported in terms of capacity, performance and latency in a very stringent manner. As a result, with data centers and infrastructure service providers aiming to maximize their revenue, and opening up their resources to several other businesses simultaneously, it becomes important to address how resource sharing can be handled in this increasingly complex scenario. Moreover, resource demands can change with time, for example, a VNF in charge of transcoding can have shifting resource demands according to the amount of multimedia in need of transcoding, which can motivate the need for migration of the VNF into a part of the infrastructure with more available resources, or closer to the user location [50]. This section addresses the dynamics involved with both resource sharing and virtual instantiations mobility.

6.4.1 Resource Sharing

When considering resource sharing in CCN, it is important to acknowledge that we are speaking of both Cloud infrastructure resources as well as data resources. The former includes compute, memory and networking resources that compose the physical infrastructure, whereas the latter refers to getting access to content generated and/or stored by other users in the same infrastructure (but potentially belonging to different virtualized instances, i.e., VMs). In this sense, resource isolation becomes an important aspect for Cloud-based service provisioning as, for example, mobile network operators may aim to operate several mobile cores, RANs and other enterprise-level deployments in the same physical substrate [38]. Typically, it is desirable for virtual resources and tenants to be strongly isolated, and unable to directly access resources outside of their domain. The use of a VM (or container) for instantiating a single service is a way to simplify resource allocation, as it improves isolation performance and meeting the demands of network services with more strict performance requirements [36]. In this way, it becomes necessary to develop administrative and operational policies that provide control capabilities to tenants in terms of defining access permissions to resources. [43] defines three models for Information And Resource Sharing (IARS), namely:

- *Invitation to Shared Project:* Where a shared project holder invites new members into the project, which is isolated from other projects and hosted inside the domain of the holder, who has full power over shared information and resources. In this sense, this model is open to the creation of trust issues, and is thus convenient only for low-assurance requirements in regards to confidentiality;
- *Joint Shared Project:* Where all collaborating participants share a project which is located in an external domain. This is also known as the OpenStack Access Control with Secure

Isolated Domain model, or OSAC-SID, where each participant has equal administration rights over resources and information, and there can be multiple isolated domains to represent individual collaborations between specific pairs of project members;

- *Shared Domain:* In a variation of the previous, this model features a single isolated domain with multiple projects for each collaboration.

In this sense, it is most important to characterize resources as well as discover them, prior to their allocation. [32] defines resource discovery as the ability of virtual network providers to discover resources available in the underlying physical infrastructure, by communicating and cooperating with infrastructure providers (i.e., assisting in optimal resource selection algorithms and processes). Resources can be further characterized by both functional requirements (i.e., type, environment, tools, operating system or network stack) and non-functional ones (i.e., performance parameters, capacity, location, cost and quality of service), with the latter varying in time. Typically, Cloud computing resource matching is also done through hash table lookups, with some enhancements such as considering geographic coordinates (for leveraging peer proximity, also known as follow-the-sun provisioning [14]) or exploiting information at the different network layers [51].

Under this prospect, sharing of resources becomes an optimization problem (in regards to a specific objective, such as maximization of remaining resources or profit) where users (i.e., services and applications) compete for VNF provisioning in the infrastructure, considering the VNF requirements, prioritization and resources available. For example, in a NFV architecture the orchestrator is the entity responsible for resource allocation [42]. [32] categorizes resource allocation approaches, complemented by [3], in the following way:

- *Centralized:* When a single entity receives and processes allocation requests, allowing for simpler, as well as more efficient, algorithms and approaches to be used, at the cost of having a single point of failure;
- *Distributed:* Aims to solve the issues of the centralized approach by involving several decision nodes for the allocation process. This allows, for example, for VNFs to be decomposed

and addressed by different decision points, but also involves synchronization communication aspects;

- *Reconfiguration:* Refers to allocation optimization as the process evolves in time, in order to maintain a more efficient usage of available resources. However, this is realized at the expense of costly migration and virtual links disruption (when they need to be reestablished);
- *Survivability:* Using additional resources available in order to cope with failures. The objective is to maximize the infrastructure provision revenue by minimizing failure penalties;
- *Redundancy:* More fault-sensitive environments can require the prior setup of backup resources to be used in a fall-back situation.

Moreover, to realize the necessary resource allocation and consequential mobility procedures, there are different challenges that need to be addressed [14], namely minimizing downtime, minimizing network reconfigurations and handling wide-area network links.

When the involved allocated resource is comprised of a virtualized network, the algorithms that instantiate it in the underlying allocated resources are named Virtual Network Embedding (VNE) [3], which encompasses both the allocation of resources in network nodes (i.e., Virtual Node Mapping) and in links (i.e., Virtual Link Mapping). As the underlying physical substrate of resources is shared, these algorithms go beyond the simple analysis of resource support of the necessary performance requirements of the virtualized networks that need to be deployed, and rather address the issue following two steps: firstly, virtual resources need to be mapped to the existing candidate physical resources and secondly, only if all resources can be mapped, the entire virtual network can then be embedded and the actual physical resources allocated.

6.4.2 Mobility Models

One of the most common mobility model aspects in virtualized environments is VM migration [52]. Realizing this procedure requires performing a copy of the disk and memory of the VM, from the old to the new location. This aspect is particularly challenging when the target location resides in a different data center (i.e., not only the migration itself causes traffic, but inter-VM traffic will have to share the inter-site connection link [50], as is in the case of migration in federated Clouds [53]). Mobility strategies for this matter can be realized under the scope of migration planning, which can consider solutions such as the realization of a pre-copy stage, compression for eliminating redundancy needs of data transmissions, jointly migrating several VMs in order to reduce overhead or migrating them in an optimal sequence or at specific scheduled times in order to avoid link bottlenecking [54]. When considering networking aspects, there is also the need to update the routing path in order to have packets destined for the VM forwarded to its new location.

Resource mobility in CCN has to consider the nature of the resource being moved. For example, NFV provides the means to orchestrate and manage different kinds of VNFs. Naturally, different VNFs have different behaviors and can impact different network system processes. For example, it is typical for VNFs to be required to be placed in the direct path of data flows [37]. Specifically in the case of virtual routers [45], despite the supportive mechanisms provided by an NFV-enabled infrastructure for moving such function (i.e., decoupling the logical functions from the underlying physical routers), it is still complex to avoid interrupting packed forwarding during the migration. This propelled the development of solutions, identified in [45], which go beyond natively running VNFs in commodity operating systems virtual machines, or even their lightweight versions such as ClickOS.[10] Concretely, FreeFlow [55] offers abstract overall system network functions to make virtual middlebox instantiation more efficient, and OpenNF [56] provides a control plane framework that is able to import and export VNF state in a coordinated fashion.

As a result, it becomes important to consider VNF placement metrics analysis, said by [37] to be: successful service mappings, total service processing times (which, according to [53], can be further broken down into VM conversion and transfer times), revenue or cost, number of active physical nodes, node buffer capacity, function processing times, function buffer demand, acceptance rate, resources utilization and traffic congestion (all under varying network conditions).

Considering the VNE perspective, resource embedding can also need to be moved elsewhere and be relocated (i.e., due to optimization

---

[10]ClickOS, http://cnp.neclab.eu/clickos/

procedures such as load balancing, or energy efficiency [57]). In this way, VNE approaches can be either static (where the infrastructure never changes) or dynamic (considering changes to both the virtual and physical infrastructure). In the latter, resource allocation maximization can involve coordination between multiple infrastructure providers, when the virtual network is spread over several physical infrastructures. In addition, [42] classifies VNE as being off-line (when all network requests are acquired and initiated beforehand) or on-line (when requests can be dynamic and have arbitrary durations).

In fact, virtual networks [33], akin to VM's, fully leverage the virtualization infrastructure capabilities and models to support their dynamic creation and modification, complemented by mobility within and between data centers. In this way, scenarios such as moving virtualized networks from loaded servers onto idle ones becomes supported. However, when moving VNs to a different substrate consider the fact that, usually, network addressing is associated with specific subnet boundaries configured within the data center (previous migration techniques enabled movement only within the same subnet [39], even though hosting facilities are increasingly being comprised of a large number of physical servers). In this sense, in order to fully support VN mobility, it is important that mechanisms be established to allow the VN to maintain critical network state (i.e., both L2 and L3 addressing) at its new target location.

Network connection maintenance in migration scenarios, within a LAN, typically resorts to the transmission of ARP advertisements of the VM's L2 and L3 addresses, prompting the local Ethernet switch to realize the necessary adjustments. In a WAN environment, other solutions are needed, such as in CloudNet [14] where Virtual Private LAN Services (VPLS) create a bridge between the original and destination site, allowing ARP messages to flow and update Ethernet switches therein, or [58] where a virtualized logical router interconnects different sites to form a virtual backplane. Other important tools to support this seamlessly are the use of SDN-based flow management and network slicing (both explored in Section 6.3 in the realization of a Virtual Overlay Network), or even a full-fledged *network hypervisor* in charge of supporting multiple VNs in the same physical substrate, using a software layer. It is then common to perceive

VN re-embedding as a costly procedure, with proposals such as [59] providing heuristic schemes to reduce the number of VNs that need re-embedding.

An example of this is the Functions Placement Problem [60] where, for each mobile core gateway, there is the need to decide whether all gateway functions can be virtualized and migrated into the data center, or if, instead, one could just host a SDN controller coupled with control-plane gateway functions. In both cases, the selected path shifts according to the function's placement in the data center, as well as from the flow control results of the SDN actions, while minimizing the total transport network load.

Finally, considering the potential for further evolving a CCN vision, connected ad hoc cloudlets [61] stand as a promising way to allow even lower communication costs and shorter transmission delay, by exploiting device-to-device (D2D) communications for computational service provisioning in mobile devices. Despite the limited resources provided by each individual device, by moving through increasing numbers of connected devices, this approach has an immense potential to provide a larger composed resource pool. This aspect, of course, brings new considerations in terms of resource mobility, with the primary being that migrations are usually done over wireless links. However, there are studies indicating that the intermittent connection associated with D2D has little impact on the optimal computing performance of a mobile cloudlet (in the long run, i.e., and considering a cloudlet formed by a very small number of hops) [62], taking advantage of the shorter link distances and the number of available surrounding devices, when compared to accessing Cloud services over long WAN latencies. In a similar way, Vehicular Cloud Computing (VCC) [63] provides a middle ground between cloudlets and data center Cloud provider solutions, as mobile nodes also usually have low computing resources, but VCC can tap into the roadside infrastructure that supports vehicular-based scenarios. VCC is also an example where local relevance of content supports the fact that short-range connections do not has as much impact on latency as thought. An example of this is Carmen [64], where the mobile connectivity of several devices belonging to the same individual is realized through CCN. Both models (i.e., cloudlet and VCC) also share the base procedures to support associated operations, namely Cloud resource discovery, Cloud formation, selection of nodes to do the processing tasks, content publishing or maintenance, Cloud maintenance and, finally, Cloud release.

6.5 Security in CCN

The Cloud computing concept has brought a wide range of advantages to the information technology (IT) area for both Cloud providers and their clients. Flexibility, scalability and availability are the most attractive assets of the Cloud computing paradigm, enabling on-demand access to computing resources that can be easily and rapidly used to provide several types of services.

This type of solution requires numerous servers to be interconnected in order to share their resources and efficiently deliver various computing services such as virtual servers, storage and databases. However, due to the introduction of new scenarios such as the Internet of Things (IoT) and Big Data, new complexity, flexibility and dynamism requirements began to surface [65]. Therefore, SDN and NFV were introduced to cope with these requirements, delivering agility, flexibility and scalability to CCN, while reducing providers' capital expenditures (CAPEX) and operating expenditures (OPEX) [66].

In this section we are going to present the main security challenges in CCN, first by explaining how security threats from traditional IT infrastructures are handled in CCN, and then by introducing new issues brought by SDN and NFV, and potential approaches to deal with them.

6.5.1 Targets and Common Network Attacks

When discussing network attacks, it is important to categorize the targets of those attacks. There are two main targets of attacks in a service: the system that runs the service and the data that it contains, receives or provides [67]. System attacks mostly affect a service's functionality or availability. The most common attacks of this kind are Denial of Service (DoS) and Distributed DoS (DDoS). These attacks are performed by making a service unavailable by interrupting or overloading its machine, which is normally carried out by making a colossal number of requests to that service. In the case of a DDoS, the attack is performed by numerous sources, e.g., botnets, with the advantages of preventing attack tracing or increasing its impact [68]. Note that in a Cloud environment, DoS can also be accomplished at the hypervisor level, when a VM tries to use all the physical resources in order to hamper other VMs from correctly functioning [69]. However, the latter is not networking related and thus is out of scope of this section. Regarding data, its integrity and confidentiality are the most common targets, where the main objective is to steal, alter or even delete the target data. Sniffing and spoofing attacks are the most common way to accomplish this. Sniffing is the process of capturing network traffic and retrieving information from it. Spoofing consists of impersonating an individual, allowing the attacker to retrieve private information, or even falsifying it.

The network attacks discussed above are well-known threats that have been around for a while in traditional networks, and there are currently common techniques to address these attacks. For example, to prevent a DoS attack, an Intrusion Detection System (IDS) may be used to detect abnormal behavior in the network or malicious hosts [70]. Spoofing attacks are typically prevented by filtering the ingress and egress traffic of the network which can be achieved using firewalls, or with more advanced techniques such as DPHCF-RTT packet filtering [71]. Regarding sniffing attacks and data protection (its integrity and confidentiality), the most common and effective method of prevention is the adoption of encryption techniques. When all the data flowing in the network is encrypted, sniffing attacks become unfeasible because without the correct encryption/decryption keys, the attacker will not be able to access the original content of that data. Protocols such as SSL, TLS and IPsec are commonly used to protect sensitive data on networks [67].

6.5.2 Security Challenges in SDN and NFV

Despite all the advantages that both SDN and NFV bring to CCN, new challenges are also introduced regarding security and privacy. First is the fact that services and networks are not running anymore on physical hardware at the user's facilities. Instead, they are running in a data center of an external entity where the provider or users can have control over them and thus can compromise them. Then, besides the existing security threats in traditional IT infrastructures,

CCN must deal with new security issues regarding the technology underlying SDN and NFV. The problem is, in CCN the data from multiple customers is flowing over shared hardware, so additional security approaches are necessary to avoid other types of attacks such as virtualization-related ones [69].

In CCN, the behavior and functionality of the networks created either using SDN or NFV must be, at least, similar to the existing traditional IT networks. Hence, Cloud networks are, inherently, subject to the already discussed common network attacks. However, in a Cloud environment there are also new security challenges derived from Cloud-specific functionalities such as network sharing, multi-tenancy, live migration and scaling.

6.5.2.1 SDN Security Challenges   SDN has potential security vulnerabilities at different points of its architecture, depicted in Figure 6.2. It is important to note that the most critical component of the whole architecture is the controller. If an attacker is able to act as the controller or gain access to an existing one, then he or she will have total control over

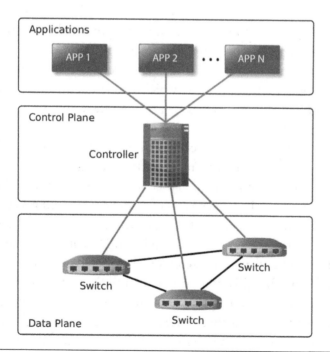

Figure 6.2   SDN architecture.

the network. Thus, security mechanisms must be considered in both the control-data and application-control interfaces. OpenFlow [72], which is one of the most popular implementations of SDN, providing a protocol for communication between the controller and the data plane devices, originally defined as a requirement that the connection between the controller and switches should be established over a TLS-encrypted connection. However, that requirement was later marked as optional on a newer version, which led many manufacturers to not support that feature [73]. This introduces a potential security issue, since an attacker can perform spoofing attacks and masquerade as a controller to gain total control over the network. As an alternative, in [74] an authentication method based on device fingerprinting is proposed for devices that are not endowed with TLS support (Figure 6.2).

Regarding the controller and interfaces that applications use to interact with it, not only encrypted connections but also authentication and authorization mechanisms must be provided. The Open Networking Foundation (ONF) proposes several security requirements for SDN controllers, derived from a threat analysis to them [75]. Moreover, in a Cloud environment, it is important allow users/applications to control only the network resources they own. Hence role-based authentication is a must, so that network resources are protected from unauthorized access. For instance, in [76] the authors propose FortNOX, a role-based authorization mechanism and security constraint enforcement for OpenFlow controllers, which is able not only to control the access to resources but also to detect rule conflicts and make decisions based on role privileges.

Additionally, the existing security threats in the traditional IT systems introduced at the beginning of this section must also be considered in SDN, both at data and control planes. In OpenFlow, for instance, the control plane consists of a single node acting as the controller. This can be seen as creating a bottleneck at the control layer and thus making the system more vulnerable to DoS attacks. One way to prevent this type of attack is to consider distributed implementations of the SDN controller [77]. Plus, distributed SDN controller implementations bring more advantages to the control plane such as redundancy, fault tolerance and scalability. Regarding the data plane, although most of the discussed techniques to address these threats can also be used in SDN networks, more advanced and SDN-aligned methods have already been presented, such as the one introduced in [78] which consists of a DDoS blocking scheme able to run on the SDN controller and making use of OpenFlow interfaces. Another interesting work in this area is the method presented in [79] which makes use of neural networks to detect DDoS attacks in a lightweight fashion.

6.5.2.2 NFV Security Challenges On the other hand, NFV introduces a considerably larger number of new security issues to CCN. Although SDN already provides a certain degree of flexibility and other advantages, the potential of NFV is far superior. However, each new resource or functionality also carries the possibility of creating new security challenges. Again, NFV provides virtualized networks very similar to the traditional ones, inheriting their threats and thus the existing solutions to cope with them can also be adopted. Hence, only NFV-specific threats will be addressed in this section.

When discussing NFV security threats, there are several points that might be possible targets of an attacker. One of them is the network infrastructure itself. Here, a malicious entity can exploit the fact that VNFs are running on shared resources, and misuse them. This can be done either by trying to take advantage of the programmability of both the Network Interface Cards (NICs) or programmable packet processors to disrupt network operations [80], or exploiting Resource Freeing Attacks (RFAs) [81], which consist essentially of modifying the workload of a victim's VM, freeing up resources for the attacker's VM. Similarly, Fraudulent Resource Consumption (FRC) [82] attacks share an identical process, but the main purpose is to make the victim incur fraudulent charges. A possible way of mitigating this kind of attack is having dedicated instances for users, and keeping a blacklist of IP addresses from which the malicious requests come, and then making regular verifications against this list [80].

Regarding virtualization, multi-tenancy allows malicious VMs to be instantiated along with victims VMs. Thus, if logical isolation is not correctly achieved, VNFs may be vulnerable to side-channel attacks. Side-channel attacks are performed by exploiting the shared resources of the VMs, e.g., CPU caches and memory, with the intent of stealing information from them. It is possible, for instance, to extract encryption keys from CPU caches when there are algorithms such as AES, DES and RSA. With this in mind, defense mechanisms against this type of attack must be considered. For instance, a system that provides VMs' locked pages in the shared caches may be used to guarantee they cannot be evicted by other VMs [83].

Additionally, live migration is also a Cloud functionality that receives attackers' focus. It is an essential asset in the Cloud world that allows VMs to be moved between servers without any down time. However, it is possible to take advantage of this migration to take control of the VM that is being migrated or instantiate one or more copies of it. This is achieved by sniffing the traffic, which contains the memory pages of the VM in the source hypervisor, and thus performing a replay attack. Hence it is necessary to consider secure methods for encrypting this traffic, in order to avoid replay or even Man-in-the-Middle (MitM) attacks. It is also advised to use hardware-based Trusted Platform Modules (TPMs) to separate all cryptographic operations from the VMs or, alternatively, other lighter and non-cryptographic methods such as Intel TxT or a Live Migration Defense Framework (LMDF) [80].

Beyond all the security vulnerabilities previously reviewed, it is also imperative that the public APIs that are provided for users interacting with the Cloud services, including networking services, also be protected. As in SDN controllers, access to these APIs must involve encryption techniques, authentication and preferably role-based authorization, in order to avoid malicious users compromising the entire infrastructure.

There is currently a huge number of active Cloud platforms around the globe, public, private and both (hybrid). Tech giants like Amazon, Alphabet, IBM, Microsoft and Oracle have been exploring Cloud business for several years, and provide their own platforms. Yet, other organizations, from small to large businesses, have been deploying their own Cloud platforms for providing their services, often adopting open source implementations. OpenStack [84] is probably the first name that comes up when discussing open source Cloud platforms. It is open source software divided into multiple components/projects that, in a distributed fashion, cooperate with each other to control a large pool of computing, storage and networking resources, and provide a Cloud platform able to serve virtual servers, containers and other resources. Regarding networking, OpenStack allows the creation of network topologies for connecting virtual servers and containers.

With the right plug-ins and frameworks, it is also possible to deliver more advanced features to OpenStack networking such as OpenFlow integration for managing SDN networks, and additional network components such as Load Balancers (LBs) and firewalls. With recent advances in both SDN and NFV and the advantages that the two provide, recommendations for them to be implemented in OpenStack have already appeared [85]. On the security front, OpenStack has some projects for solving particular issues. For instance, the Anchor project provides a Public Key Infrastructure (PKI) for managing client certificates with a configurable validity period. It introduces the concept of passive revocation, which consists of revoking client certificates by simply not re-issuing them, a process that enables quick certification revocation/rotation when used with short lifetimes [86].

Security in CCN has been an interesting research challenge in recent years, both in academia and industry. People search for the Cloud's flexibility and scalability, but need the robustness, security and performance of the physical world. This triggers a lot of research activities towards the development of solutions that take this into consideration. Although both SDN and NFV share some basic principles, they differ in concept. Yet they have been brought together over time and some research considering both concepts has been presented, combining their advantages for deployment in diverse areas such as the IoT [87], Virtual Optical Networks [88] and satellite communications in 5G networks [89]. In fact, both concepts will have an important role in 5G networks as the two allow reducing both CAPEX and OPEX, while increasing revenue and providing various key functionalities, essentially dynamic deployment and scaling of functions in Cloud networking [90]. Inevitably, 5G networks will raise particular security concerns [91] leading to the rise of new efforts towards security standardization and architectures in 5G such as 5G-ENSURE [92], a European project whose main purpose is to develop a 5G Reference Architecture with security and privacy in mind.

Acknowledgments

This work is funded by FCT/MEC through national funds and when applicable co-funded by the FEDER PT2020 partnership agreement under the projects UID/EEA/50008/2013 and PTDC/

EEI-TEL/30685/2017, and by the Integrated Programme of SR&TD SOCA (Ref. CENTRO-01-0145-FEDER-000010) and co-funded by Centro 2020 program, Portugal 2020, European Union, through the European Regional Development Fund.

Acronyms

AAA	Authentication, Authorization and Accounting
API	Application Program Interface
ARPU	Average Revenue Per User
BGP	Border Gateway Protocol
CAPEX	Capital Expenditure
CCN	Cloud-Centric Computing
CDN	Content Delivery Network
D2	Device to Device
DDoS	Distributed Denial of Service
DoS	Denial of Service
E2E	End-to-end
FRC	Fraudulent Resource Consumption
IaaS	Infrastructure-as-a-Service
IARS	Information And Resource Sharing
ICT	Information and Communication Technologies
IDS	Intrusion and Detection System
IMS	IP Multimedia Subsystem
IoT	Internet of Things
IT	Information Technology
InP	Infrastructure Provider
LMDF	Live Migration Defense Framework
MANO	Management and Orchestration
MEC	Mobile Edge Computing
MNO	Mobile Network Operator
NaaS	Network-as-a-Service
NFV	Network Function Virtualization
NFVI	Network Function Virtualization Infrastructure
NFVO	Network Function Virtualization Orchestrator
NGN	Next Generation Network
NGSON	Next Generation Service Overlay Network
NIC	Network Interface Card

NVGRE	Network Virtualization using Generic Routing ENcapsulation
OSAC-SID	OpenStack Access Control with Secure Isolated Domain
OPEX	Operational Expenditure
PaaS	Platform-as-a-Service
PKI	Public Key Infrastructure
QoE	Quality of Experience
QoS	Quality of Service
RFA	Resource Freeing Attacks
SaaS	Software-as-a-Service
SDN	Software Defined Networking
SFC	Service Function Chaining
SLA	Service Level Agreement
SP	Service Provider
STT	Stateless Transport Tunneling
TPM	Trusted Platform Modules
VIM	Virtualized Infrastructure Manager
VCC	Vehicular Cloud Computing
VLAN	Virtual Local Area Network
VNE	Virtual Network Embedding
VON	Virtual Overlay Network
VPLS	Virtual Private LAN Services
VPN	Virtual Private Network
VxLAN	Virtual Extensible LAN
VM	Virtual Machine
VNF	Virtual Network Function

References

1. Y. Zhou, X. Yang, Y. Li, D. Jin, L. Su, and L. Zeng. Incremental re-embedding scheme for evolving virtual network requests. *IEEE Communications Letters*, 17(5):1016–1019, 2013.
2. H. Wang, Y. Li, Y. Zhang, and D. Jin. Virtual machine migration planning in software-defined networks. In *2015 IEEE Conference on Computer Communications (INFOCOM)*, pp. 487–495, 2015.
3. T. Wood, K. K. Ramakrishnan, P. Shenoy, J. Van der Merwe, J. Hwang, G. Liu, and L. Chaufournier. Cloudnet: Dynamic pooling of Cloud resources by live WAN migration of virtual machines. *IEEE/ACM Transactions on Networking*, 23(5):1568–1583, 2015.

4. D. V. Bernardo and B. B. Chua. Introduction and analysis of SDN and NFV security architecture (SN-SECA). In *Advanced Information Networking and Applications (AINA), 2015 IEEE 29th International Conference on*, pp. 796–801. IEEE, 2015.

5. 5G-ENSURE. http://www.5gensure.eu/. Accessed: 2017-08-09.

6. I. D. Cardoso, J. P. Barraca, C. Gonalves, and R. L. Aguiar. Seamless integration of Cloud and fog networks. *International Journal of Network Management*, 26(6):435–460, 2016.

7. T. Kim, M. Peinado, and G. Mainar-Ruiz. STEALTHMEM: System-level protection against cache-based side channel attacks in the cloud. In *USENIX Security Symposium*, pp. 189–204, 2012.

8. J. Liu, Y. Li, D. Jin, L. Su, and L. Zeng. Traffic aware cross-site virtual machine migration in future mobile Cloud computing. *Mobile Networks and Applications*, 20(1):62–71, 2015.

9. A. Gember-Jacobson, R. Viswanathan, C. Prakash, R. Grandl, J. Khalid, S. Das, and A. Akella. Opennf: Enabling innovation in network function control. In *Proceedings of the 2014 ACM Conference on SIGCOMM*, SIGCOMM'14, pp. 163–174, New York, NY, 2014. ACM.

10. F. Hao, T. V. Lakshman, S. Mukherjee, and H. Song. Enhancing dynamic Cloud-based services using network virtualization. *SIGCOMM Computer Communication Review*, 40(1):67–74, 2010.

11. A. Belbekkouche, M. M. Hasan, and A. Karmouch. Resource discovery and allocation in network virtualization. *IEEE Communications Surveys Tutorials*, 14(4):1114–1128, Fourth 2012.

12. J. Qadir, N. Ahmed, F. Z. Yousaf, and A. Taqweem. Network as a service: The new vista of opportunities. *CoRR*, abs/1606.03060, 2016.

13. M. Alaluna, Fernando M. V. Ramos, and Nuno Neves. (Literally) above the Clouds: virtualizing the network over multiple Clouds. *CoRR*, abs/1512.01196, 2015.

14. M. S. Carmo, S. Jardim, T. de Souza, A. V. Neto, R. Aguiar, and D. Corujo. Towards enhanced connectivity through WLAN slicing. In *2017 Wireless Telecommunications Symposium (WTS)*, pp. 1–7, April 2017.

15. B. Sousa, L. Cordeiro, P. Simes, A. Edmonds, S. Ruiz, G. A. Carella, M. Corici, et al. Toward a fully cloudified mobile network infrastructure. *IEEE Transactions on Network and Service Management*, 13(3):547–563, 2016.

16. Anchor. https://wiki.openstack.org/wiki/Security/Projects/Anchor. Accessed: 2017-08-10.

17. ITU. Future networks, with focus on IMT-2020, Cloud computing and trusted network infrastructures. In *ITU SG13*, 2017.

18. M. Y. Luo and J. Y. Chen. Virtual transits: A flexible platform for network virtualization across data centers. In *2014 IEEE 6th International Conference on Cloud Computing Technology and Science*, pp. 563–570, 2014.

19. A. Dixit, F. Hao, S. Mukherjee, T. V. Lakshman, and R. Kompella. Towards an elastic distributed SDN controller. In *ACM SIGCOMM Computer Communication Review*, volume 43, pp. 7–12. ACM, 2013.

20. Network Functions Virtualisation: White Paper #3 ETSI NFV ISG. https://portal.etsi.org/NFV/NFV_White_Paper3.pdf, January 2015.
21. M. Ersue. ETSI NFV management and orchestration: An overview. A presentation for IETF #88, Vancouver, Canada, https://www.ietf.org/proceedings/88/slides/slides-88-opsawg-6.pdf [online, accessed 12-Dec 2018].
22. B. A. A. Nunes, M. Mendonca, X. N. Nguyen, K. Obraczka, and T. Turletti. A survey of software-defined networking: Past, present, and future of programmable networks. *IEEE Communications Surveys Tutorials*, 16(3):1617–1634, 2014.
23. K. S. Arulmozhi, R. Karthikeyan, and B. Chandra Mohan. *Optimizing Resource Sharing in Cloud Computing*, pp. 50–55. Springer, Berlin, Germany, 2011.
24. R. Jain and S. Paul. Network virtualization and software defined networking for Cloud computing: A survey. *IEEE Communications Magazine*, 51(11):24–31, 2013.
25. S. Seshan, M. Stemm, and R. H. Katz. Spand: Shared passive network performance discovery. In *Proceedings of the USENIX Symposium on Internet Technologies and Systems on USENIX Symposium on Internet Technologies and Systems*, USITS'97, pp. 13–13, Berkeley, CA, 1997. USENIX Association.
26. E. Trouva et al. Tnova d2.22 overall system. architecture and interfaces v1.0. In *FP7 T-NOVA: Network Functions As-A-Service over Virtualised Infrastructures*, 2015.
27. J. Costa-Requena, J. L. Santos, V. F. Guasch, K. Ahokas, G. Premsankar, S. Luukkainen, O. López Pérez, M. U. Itzazelaia, I. Ahmad, M. Liyanage, et al. SDN and NFV integration in generalized mobile network architecture. In *Networks and Communications (EuCNC), 2015 European Conference on*, pp. 154–158. IEEE, 2015.
28. A. Wolke, M. Bichler, F. Chirigati, and V. Steeves. Reproducible experiments on dynamic resource allocation in Cloud data centers. *Information Systems*, 59:98–101, 2016.
29. W. Stallings. *Foundations of Modern Networking: SDN, NFV, QoE, IoT, and Cloud*. Addison-Wesley Professional, 2015.
30. P. K. Agyapong, M. Iwamura, D. Staehle, W. Kiess, and A. Benjebbour. Design considerations for a 5G network architecture. *IEEE Communications Magazine*, 52(11):65–75, 2014.
31. J. Gil Herrera and J. F. Botero. Resource allocation in NFV: A comprehensive survey. *IEEE Transactions on Network and Service Management*, 13(3):518–532, 2016.
32. Y. Li and W. Wang. Can mobile cloudlets support mobile applications? In *IEEE INFOCOM 2014: IEEE Conference on Computer Communications*, pp. 1060–1068, 2014.
33. S. Rajagopalan, D. Williams, H. Jamjoom, and A. Warfield. Split/merge: System support for elastic execution in virtual middleboxes. In *Presented as Part of the 10th USENIX Symposium on Networked Systems Design and Implementation (NSDI 13)*, pages 227–240, Lombard, IL, 2013. USENIX.

34. A. Greenhalgh, F. Huici, M. Hoerdt, P. Papadimitriou, M. Handley, and L. Mathy. Flow processing and the rise of commodity network hardware. *SIGCOMM Computer Communication Review*, 39(2):20–26, 2009.

35. P. Bellavista. Mobile Cloud networking: Lessons learnt, open research directions, and industrial innovation opportunities. In *2016 4th IEEE International Conference on Mobile Cloud Computing, Services, and Engineering (MobileCloud)*, pages 79–80, March 2016.

36. X. Foukas, G. Patounas, A. Elmokashfi, and M. K. Marina. Network slicing in 5G: Survey and challenges. *IEEE Communications Magazine*, 55(5):94–100, 2017.

37. ETSI European Telecommunications Standard Institute. Architectural framework; GS NFV 002 v1.2.1. In *Network Functions Virtualisation*, 2014.

38. S. Lim, J. Ha, H. Kim, Y. Kim, and S. Yang. A SDN-oriented DDoS blocking scheme for botnet-based attacks. In *Ubiquitous and Future Networks (ICUFN), 2014 Sixth International Conf on*, pp. 63–68. IEEE, 2014.

39. N. Grozev and R. Buyya. Inter-Cloud architectures and application brokering: Taxonomy and survey. *Software: Practice and Experience*, 44(3):369–390, 2014.

40. R. Muñoz, R. Vilalta, R. Casellas, R. Martnez, T. Szyrkowiec, A. Autenrieth, V. López, and D. López. SDN/NFV orchestration for dynamic deployment of virtual SDN controllers as VNF for multi-tenant optical networks. In *Optical Fiber Communications Conference and Exhibition (OFC), 2015*, pp. 1–3. IEEE, 2015.

41. B. Han, V. Gopalakrishnan, L. Ji, and S. Lee. Network function virtualization: Challenges and opportunities for innovations. *IEEE Communications Magazine*, 53(2):90–97, 2015.

42. J. D. Guyton and M. F. Schwartz. Locating nearby copies of replicated internet servers. In *Proceedings of the Conference on Applications, Technologies, Architectures, and Protocols for Computer Communication*, SIGCOMM'95, pages 288–298, New York, 1995. ACM.

43. X. Tang and J. Xu. Qos-aware replica placement for content distribution. *IEEE Transactions on Parallel and Distributed Systems*, 16(10):921–932, 2005.

44. H. T. Dinh, C. Lee, D. Niyato, and P. Wang. A survey of mobile Cloud computing: architecture, applications, and approaches. *Wireless Communications and Mobile Computing*, 13(18):1587–1611, 2013.

45. M. Chen, Y. Hao, Y. Li, C. F. Lai, and D. Wu. On the computation offloading at ad hoc cloudlet: Architecture and service modes. *IEEE Communications Magazine*, 53(6):18–24, 2015.

46. L. Coppolino, S. DAntonio, G. Mazzeo, and L. Romano. Cloud security: Emerging threats and current solutions. *Computers & Electrical Engineering*, 59:126–140, 2017.

47. M. Al-Fares, A. Loukissas, and A. Vahdat. A scalable, commodity data center network architecture. *SIGCOMM Comput. Commun. Rev.*, 38(4):63–74, August 2008.

48. Y. Ai, M. Peng, and K. Zhang. Edge Cloud computing technologies for internet of things: A primer. *Digital Communications and Networks*, 2017.

49. V. Varadarajan, T. Kooburat, B. Farley, T. Ristenpart, and M. M. Swift. Resource-freeing attacks: Improve your Cloud performance (at your neighbor's expense). In *Proceedings of the 2012 ACM Conference on Computer and Communications Security*, pp. 281–292. ACM, 2012.

50. N. McKeown, T. Anderson, H. Balakrishnan, G. Parulkar, L. Peterson, J. Rexford, S. Shenker, and J. Turner. OpenFlow: Enabling innovation in campus networks. *ACM SIGCOMM Computer Communication Review*, 38(2):69–74, 2008.

51. T. X. Tran, A. Hajisami, P. Pandey, and D. Pompili. Collaborative mobile edge computing in 5g networks: New paradigms, scenarios, and challenges. *IEEE Communications Magazine*, 55(4):54–61, 2017.

52. R. Ferrús, H. Koumaras, O. Sallent, G. Agapiou, T. Rasheed, M.-A. Kourtis, C. Boustie, P. Gélard, and T. Ahmed. SDN/NFV-enabled satellite communications networks: Opportunities, scenarios and challenges. *Physical Communication*, 18:95–112, 2016.

53. Q. Duan, Y. Yan, and A. V. Vasilakos. A survey on service-oriented network virtualization toward convergence of networking and Cloud computing. *IEEE Transactions on Network and Service Management*, 9(4):373–392, December 2012.

54. M. A. Marotta, N. Kaminski, I. Gomez-Miguelez, L. Z. Granville, J. Rochol, L. DaSilva, and C. B. Both. Resource sharing in heterogeneous Cloud radio access networks. *IEEE Wireless Communications*, 22(3):74–82, 2015.

55. IEEE. Ieee standard for the functional architecture of next generation service overlay network. In *IEEE SA1903-2011*, 2011.

56. R. Maheshwari, C. Rama Krishna, and M. Sridhar Brahma. Defending network system against IP spoofing based distributed DoS attacks using DPHCF-RTT packet filtering technique. In *Issues and Challenges in Intelligent Computing Techniques (ICICT), 2014 International Conference on*, pp. 206–209. IEEE, 2014.

57. B. Pfaff, J. Pettit, T. Koponen, K. Amidon, M. Casado, and S. Shenker. Extending networking into the virtualization layer. In *Proceedings of Workshop on Hot Topics in Networks (HotNets-VIII)*, 2009.

58. J. Soares, C. Gonalves, B. Parreira, P. Tavares, J. Carapinha, J. P. Barraca, R. L. Aguiar, and S. Sargento. Toward a telco Cloud environment for service functions. *IEEE Communications Magazine*, 53(2):98–106, 2015.

59. N. Laoutaris, V. Zissimopoulos, and I. Stavrakakis. On the optimization of storage capacity allocation for content distribution. *Computer Networks*, 47(3):409–428, 2005.

60. K.-H. Kim, S.-J. Lee, and P. Congdon. On Cloud-centric network architecture for multi-dimensional mobility. In *Proceedings of the First Edition of the MCC Workshop on Mobile Cloud Computing*, MCC'12, pp. 1–6, New York, 2012. ACM.

61. R. Braga, E. Mota, and A. Passito. Lightweight DDoS flooding attack detection using NOX/OpenFlow. In *Local Computer Networks (LCN), 2010 IEEE 35th Conference on*, pp. 408–415. IEEE, 2010.

62. P. Porras, S. Shin, V. Yegneswaran, M. Fong, M. Tyson, and G. Gu. A security enforcement kernel for OpenFlow networks. In *Proceedings of the First Workshop on Hot Topics in Software Defined Networks*, pp. 121–126. ACM, 2012.

63. C. Kenyon and M. Baker. Making the most of SDN and NFV with Ubuntu OpenStack, 2015.

64. J. Qadir, N. Ahmed, and N. Ahad. Building programmable wireless networks: An architectural survey. *EURASIP Journal on Wireless Communications and Networking*, 2014(1) 172, 2014.

65. E. Hernandez-Valencia, S. Izzo, and B. Polonsky. How will NFV/SDN transform service provider opex? *IEEE Network*, 29(3):60–67, 2015.

66. A. Welzel, C. Rossow, and H. Bos. On measuring the impact of DDoS botnets. In *Proceedings of the Seventh European Workshop on System Security*, p. 3. ACM, 2014.

67. A. Cimmino and P. Donadio. D4.2: ETICS architecture and functional entities high level design. In *ETICS Economics And Technologies for Inter Carrier Services*, 2011.

68. L. Wang, F. Zhang, A. V. Vasilakos, C. Hou, and Z. Liu. Joint virtual machine assignment and traffic engineering for green data center networks. *SIGMETRICS Performance Evaluation Review*, 41(3):107–112, 2014.

69. P. Schneider and G. Horn. Towards 5G security. In *Trustcom/BigDataSE/ISPA, 2015 IEEE*, volume 1, pp. 1165–1170. IEEE, 2015.

70. H. Hawilo, A. Shami, M. Mirahmadi, and R. Asal. NFV: State of the art, challenges, and implementation in next generation mobile networks (VEPC). *IEEE Network*, 28(6):18–26, 2014.

71. O. Sefraoui, M. Aissaoui, and M. Eleuldj. OpenStack: Toward an open-source solution for Cloud computing. *International Journal of Computer Applications*, 55(3), 2012.

72. J. Idziorek, M. Tannian, and D. Jacobson. Attribution of fraudulent resource consumption in the Cloud. In *Cloud Computing (CLOUD), 2012 IEEE 5th International Conference on*, pp. 99–106. IEEE, 2012.

73. N. Gray, T. Zinner, and P. Tran-Gia. Enhancing SDN security by device fingerprinting. In *Integrated Network and Service Management (IM), 2017 IFIP/IEEE Symposium on*, pp. 879–880. IEEE, 2017.

74. Y. Li and M. Chen. Software-defined network function virtualization: A survey. *IEEE Access*, 3:2542–2553, 2015.

75. Y. Xue and R. Deters. Resource sharing in mobile Cloud-computing with COAP. *Procedia Computer Science*, 63:96–103, 2015. The 6th International Conference on Emerging Ubiquitous Systems and Pervasive Networks (EUSPN 2015)/The 5th International Conference on Current and Future Trends of Information and Communication Technologies in Healthcare (ICTH-2015)/Affiliated Workshops.

76. H.-Y. Tsai, M. Siebenhaar, A. Miede, Y. Huang, and R. Steinmetz. Threat as a service? Virtualizations impact on Cloud security. *IT professional*, 14(1):32–37, 2012.
77. A. Fischer, J. F. Botero, M. T. Beck, H. de Meer, and X. Hesselbach. Virtual network embedding: A survey. *IEEE Communications Surveys Tutorials*, 15(4):1888–1906, 2013.
78. A. Beloglazov, J. Abawajy, and R. Buyya. Energy-aware resource allocation heuristics for efficient management of data centers for Cloud computing. *Future Generation Computer Systems*, 28(5):755–768, 2012.
79. J. Hwang, K. K. Ramakrishnan, and T. Wood. Netvm: High performance and flexible networking using virtualization on commodity platforms. In *11th USENIX Symposium on Networked Systems Design and Implementation (NSDI 14)*, pp. 445–458, Seattle, WA, 2014. USENIX Association.
80. F. Meneses, D. Corujo, C. Guimaraes, and R. Aguiar. An abstraction framework for flow mobility in multi-technology 5G environments using virtualization and SDN. In *Netsoft 2017*, 2017.
81. S. Latre, J. Famaey, F. D. Turck, and P. Demeester. The fluid Internet: Service-centric management of a virtualized future Internet. *IEEE Communications Magazine*, 52(1):140–148, 2014.
82. M. D. Firoozjaei, J. Paul Jeong, H. Ko, and H. Kim. Security challenges with network functions virtualization. *Future Generation Computer Systems*, 67:315–324, 2017.
83. K. Benton, L. Jean Camp, and C. Small. Openflow vulnerability assessment. In *Proceedings of the Second ACM SIGCOMM Workshop on Hot Topics in Software Defined Networking*, pp. 151–152. ACM, 2013.
84. ONF. Security foundation requirements for SDN controllers, 2016.
85. T. Taleb and A. Ksentini. Follow me Cloud: Interworking federated Clouds and distributed mobile networks. *IEEE Network*, 27(5):12–19, 2013.
86. E. Lee, E. K. Lee, M. Gerla, and S. Y. Oh. Vehicular Cloud networking: Architecture and design principles. *IEEE Communications Magazine*, 52(2):148–155, 2014.
87. H.-J. Liao, C.-H. Richard Lin, Y.-C. Lin, and K.-Y. Tung. Intrusion detection system: A comprehensive review. *Journal of Network and Computer Applications*, 36(1):16–24, 2013.
88. N. Omnes, M. Bouillon, G. Fromentoux, and O. Le Grand. A programmable and virtualized network & IT infrastructure for the internet of things: How can NFV & SDN help for facing the upcoming challenges? In *Intelligence in Next Generation Networks (ICIN), 2015 18th International Conference on*, pp. 64–69. IEEE, 2015.
89. A. Basta, W. Kellerer, M. Hoffmann, H. J. Morper, and K. Hoffmann. Applying NFV and SDN to LTE mobile core gateways, the functions placement problem. In *Proceedings of the 4th Workshop on All Things Cellular: Operations, Applications, Challenges*, AllThingsCellular'14, pp. 33–38, New York, 2014. ACM.

90. D. Bernstein and Y. Demchenko. The IEEE intercloud testbed: Creating the global Cloud of Clouds. In *2013 IEEE 5th International Conference on Cloud Computing Technology and Science*, volume 2, pp. 45–50, 2013.

91. S. Kchele, C. Spann, F. J. Hauck, and J. Domaschka. Beyond IaaS and PaaS: An extended Cloud taxonomy for computation, storage and networking. In *2013 IEEE/ACM 6th International Conference on Utility and Cloud Computing*, pp. 75–82, 2013.

92. Jiaqiang Liu, Yong Li, and Depeng Jin. SDN-based live VM migration across data centers. In *Proceedings of the 2014 ACM Conference on SIGCOMM*, SIGCOMM'14, pages 583–584, New York, 2014. ACM.

Abbreviations

5G	5th-generation wireless systems
6LoWPAN	IPv6 over low-power wireless personal area networks
AAA	authentication, authorization and accounting
ACO	ant colony optimization
AH	authentication header
AHS	adaptive HTTP streaming
AP	access point
API	application program interface
ARPU	average revenue per user
AVC	advanced video coding
BGP	border gateway protocol
BS	base station
CAPEX	capital expenditure
CCN	content-centric networking
CCN	cloud-centric computing
CDN	content delivery networks
CDNaaS	content delivery network as a service
CH	cluster head
CIF	common intermediate format
CNS	centric-based networking systems
CoA	care of addresses
CoAP	constrained application protocol

CR	content routers
CS	content store
CSI	channel state information
D2D	device-to-device
DASH	dynamic adaptive streaming over HTTP
DCN	device-centric network
DDoS	distributed denial of service
DHT	distributed hash tables
DLNA	digital living network alliance
DMCA	digital millennium copyright act
DNS	domain name system
DoF	degrees of freedom
DONA	data-oriented network architecture
DoS	denial of service
DRM	digital rights management
E2E	end-to-end
EID	EntityID
EID	endpoint identifier
ELOC	entity locators
ERDF	European regional development fund
ESP	encapsulating security payload
EU FP6	European Union/European Commission sixth framework programme
FI	future internet
FIB	forwarding information base
FIXP	future internet eXchange point
FPA	framework partnership agreement
FPS	frames per second
FRC	fraudulent resource consumption
GEM	graph EMbedding for sensor networks
H-DHT	hierarchical distributed hash tables
HEVC	high efficiency video coding
HIMALIS	heterogeneity inclusion and mobility adaptation through locator ID separation
HIP	host identity protocol
HLS	HTTP live streaming
HTML	hypertext markup language
HTTP	hypertext transfer protocol

I/O	input/output
IaaS	infrastructure as a service
IARS	information and resource sharing
ICN	information-centric networking
ICNRG	information-centric networking research group
ICT	information and communication technologies
ID	identification
IDS	intrusion and detection system
ILNP	identifier-locator network protocol
IMS	IP multimedia subsystem
InP	infrastructure provider
IoT	internet of things
IP	internet protocol
IPsec	internet protocol security
IPv4	internet protocol version 4
IPv6	internet protocol version 6
IRTF	internet research task force
ISDN	integrated services digital network
IT	information technology
kbps	kilobit per second
LISP	location/ID separation protocol
LLN	low power and lossy networks
LMDF	live migration defense framework
LPWAN	low-power wide-area network
LTE	long-term evolution
M2M	machine-to-machine
MAC	media access control
MANO	management and orchestration
MCS	mobile content source
MEC	mobile edge computing
MNO	mobile network operator
MOFI	mobile oriented future internet
MPD	media presentation description
MPEG	moving picture experts group
MQTT	message queuing telemetry transport
MTU	maximum transmission unit
MVNO	mobile virtual network operator
NaaS	network as a service

NACK	negative acknowledgement
NALU	network abstraction layer unit
NAND	NOT AND
NCN	network-centric networking
NDN	named data networking
NDO	named data object
NEPI	network experimentation programming interface
NetInf	network of information
NFV	network function virtualization
NFVI	network function virtualization infrastructure
NFVO	network function virtualization orchestrator
NGN	next generation network
NGSON	next generation service overlay network
NIC	network interface card
NICT	National Institute of Information and Communication Technology of Japan
NodeID	node identity internetworking architecture
NOR	NOT OR
NPSN	named publish subscribe networking
NVGRE	network virtualization using generic routing encapsulation
O-ICN	overlay ICN
OIDC	openID connect
OPEX	operational expenditure
OSAC-SID	OpenStack access control with secure isolated domain
OSE	object-oriented service entities
OSI	open system interconnect
OTT	over-the-top
P2P	peer-to-peer
PaaS	platform as a service
PAN	personal area network
PIT	pending interest table
PKI	public key infrastructure
POSIX	portable operating system interface
PSIRP	publish-subscribe internet routing paradigm
PSNR	peak signal noise ratio
PURSUIT	publish-subscribe internet technology
QoE	quality of experience

QoS	Quality of Service
QP	quantization parameter
RANGI	routing architecture for the next generation internet
RFA	resource freeing attacks
RFC	request for comments
RFID	radio frequency identification
RLOCs	routing locators
RPL	IPv6 routing protocol for LLNs
RTMP	real-time messaging protocol
RTP	real-time transport protocol
RTSP	real-time streaming protocol
RTT	round trip time
SaaS	software as a service
SAP	session announcement protocol
SCN	storage-centric networking
SCTP	stream control transmission protocol
SDN	software-defined networking
SDP	service discovery protocol
SDP	session description protocol
SDS	software-defined storage systems
SFC	service function chaining
SLA	service level agreement
SMARTFIRE	Enabling SDN ExperiMentAtion in WiReless Testbeds
SNR	signal-to-noise ratio
SP	service provider
SrCN	service-centric networking
SSL	secure sockets layer
STT	stateless transport tunneling
SVC	scalable video coding
TCP	transmission control protocol
TinyPEDS	tiny persistent encrypted data storage
TPM	trusted platform modules
UCN	user-centric networking
UDP	user datagram protocol
UMTS	universal mobile telecommunications service
UPnP	universal plug and play

URI	uniform resource identifiers
URL	uniform resource locator
UUID	universally unique identifier
VBR	variable bit-rate
VCC	vehicular cloud computing
VCEG	video coding experts group
VCR	video cassette recorder
VIM	virtualized infrastructure manager
VLAN	virtual local area network
VM	virtual machine
VNE	virtual network embedding
VNF	virtual network function
VoD	video on demand
VON	virtual overlay network
VPLS	virtual private LAN services
VPN	virtual private network
VxLAN	virtual extensible LAN
WBAN	wireless body area network
WiMAX	worldwide interoperability for microwave access
WLAN	wireless local area network
WPAN	wireless personal area network
WSN	wireless sensor network
WWW	world wide web
XML	eXtensible markup language

Index